RADIO ANTENNAS
AND PROPAGATION

RADIO ANTENNAS AND PROPAGATION

WILLIAM GOSLING

Newnes

OXFORD BOSTON JOHANNESBURG MELBOURNE NEW DELHI SINGAPORE

Newnes
An imprint of Butterworth-Heinemann
Linacre House, Jordan Hill, Oxford OX2 8DP
225 Wildwood Avenue, Woburn, MA 01801-2041
A division of Reed Educational and Professional Publishing Ltd

 A member of the Reed Elsevier plc group

First published 1998
Transferred to digital printing 2004
© William Gosling 1998

All rights reserved. No part of this publication may be
reproduced in any material form (including photocopying
or storing in any medium by electronic means and whether
or not transiently or incidentally to some other use of
this publication) without the written permission of the
copyright holder except in accordance with the provisions
of the Copyright, Designs and Patents Act 1988 or under
the terms of a licence issued by the Copyright Licensing
Agency Ltd, 90 Tottenham Court Rd, London, England W1P 9HE.
Applications for the copyright holder's written permission
to reproduce any part of this publication should be
addressed to the publishers

British Library Cataloguing in Publication Data
A catalogue record for this book is available from the British Library

ISBN 0 7506 3741 2

Library of Congress Cataloging in Publication Data
A catalogue record for this book is available from the Library of Congress

Typeset by David Gregson Associates, Beccles, Suffolk

Contents

Preface	vii
1 Introduction	1
Part One: Antennas	**17**
2 Antennas: getting started	19
3 The inescapable dipole	26
4 Antenna arrays	52
5 Parasitic arrays	72
6 Antennas using conducting surfaces	81
7 Wide-band antennas	103
8 Odds and ends	116
9 Microwave antennas	133
Part Two: Propagation	**151**
10 Elements of propagation	153
11 The atmosphere	165
12 At ground level	177
13 The long haul	202
Appendix: Feeders	244
Further reading	254
Index	**255**

PREFACE

Textbooks on radio antennas and propagation have changed little over the last 50 years. Invariably they base themselves on the famous electromagnetic equations described by James Clerk Maxwell, a great nineteenth-century genius of theoretical physics (Torrance, 1982). Maxwell's equations brilliantly encompassed all the electromagnetic phenomena known by his time (except photoelectric long-wave cut-off, which remained a mystery). To this day, the classic textbooks on antennas and propagation treat the subject as a series of solutions of Maxwell's equations fitted to practical situations. Doing this turns out to be far from easy in all but a very few cases. Even so, by ingenuity and approximation, solutions are revealed which correspond quite well to what may be observed and measured in real life.

Maxwell's equations work; they did when he announced them and they still do. As applicable mathematics they remain a valid and valuable tool. Nevertheless, the physics he used to derive them is entirely discredited. Maxwell based his electromagnetics on the notion of forces and waves acting in a universal elastic medium called the ether. Invisible and impalpable, it nevertheless permeated the whole universe. Yet only six years after his death the famous Michelson–Morley experiments began to cast doubt on the existence of the ether. Now the idea is dead, thanks to the universal adoption of relativistic physics and quantum theory. In our present-day interpretation, radio energy consists of photons, electromagnetic quanta which are incredibly small and strange, particles that also have wave properties.

Quantum mechanics, because of the very oddness of some of its predictions, has been subjected to the most rigorous processes of experimental testing conceivable, more so than any other branch of physics. One day things might change, but for now and the foreseeable future, quantum theory is the most firmly established of all physical ideas. Yet for half a century we have gone on teaching electromagnetics to generations of engineering students as if the quantum revolution had never happened. Why so?

It is true that the classical Maxwell approach does provide a good mathematical model of electromagnetic phenomena. Nowhere in radio engineering does it blatantly fail, as it does in optics and spectroscopy. The radio frequency quantum is much less energetic than its optical counterpart, so any detectable energy involves very many of them. As a result, effects attributable to individuals are not seen, and everything averages out to the classical picture. So if radio engineers ignore quantum mechanics nothing actually goes wrong for them, and this was long thought reason enough for leaving it out of books and courses. It seemed an unnecessary complication.

Times change, however. Modern electrical engineering students must pick their way through some quantum mechanics to understand semiconductor devices; it is no longer an optional extra. But to use quantum explanations about transistors and microcircuits yet ignore them when it comes to radio destroys the natural unity of our subject, fails to make important connections and seems arbitrary. Besides, 'difficult' ideas grow easier with use and a quantum orientation to radio no longer makes the subject less accessible to modern students. On the contrary, sometimes quantum notions give an easier insight than the old classical approach. The 'feel' is so much less abstract, so much more real-world oriented. Anyway, I cannot help believing that we ought to teach our students the best we know, particularly since we have no idea what will be important to them in the future. So, start to finish, this book takes an approachable but persistently quantum-oriented stance, and in my mind that is what justified writing it. My hope is that it will encourage those who have long wanted to teach the subject in a more modern way.

As to acknowledgements, first my undying gratitude to generations of final year students at the University of Bath, from whom I discovered how best to teach this subject. Heartfelt thanks also to Duncan Enright at Newnes, for encouraging me to turn the course into a book.

William Gosling

CHAPTER 1

1 INTRODUCTION

This book is about how radio energy is released (**transmission**), how it moves from one place to another (**propagation**) and how it is captured again (**reception**). Understanding all this is indispensable for communications engineers because during the twentieth century radio has become a supremely important means of carrying information.

First used by ships at sea, soon after 1900, radio systems were quickly developed for broadcasting (sound from around 1920, television after 1936). At much the same time came air traffic control, emergency services (police, fire, ambulance) and later private mobile radio, with users ranging from taxi drivers in the city streets to civil engineers on major construction projects. The military were enthusiastic users of radio from the start, notably for battlefield communication (especially in tanks), for warships, both surface and submarine, and the command and control of military aircraft. In the second quarter of the twentieth century radio navigation systems, which enabled ships and aircraft to obtain accurate 'fixes' on their position, spread to give worldwide coverage. A modification of standard radio techniques, permitting reception of reflected energy, led (from about 1938) to the extensive use of radar for the detection, and later even imaging, of distant objects such as ships, aircraft or vehicles.

The worldwide annual turnover of the radio industry (in all its many forms) still exceeds that of the computer industry, and it is growing just as fast. In recent times, optical fibres have replaced

2 Radio Antennas and Propagation

radio, to some extent, for communication between fixed locations, but for all situations in which one or both ends of a communication link may be mobile or subject to movement, radio remains the only information-bearer technology. Early radio engineers struggled to get the maximum possible range from their systems, but today, as well as continued interest in long ranges, there is also an explosive growth in the use of short-range radio systems. Cellular radio telephones are the most obvious example. Short-range radio has the important advantage that it enables more users to be accommodated in the same radio bands without interfering with each other.

All of this explosive technological development depends on the transmission, propagation and reception of radio energy. So what is radio energy?

1.1 What radio energy really is

Radio energy is similar to light. It propagates freely in space as a stream of very small, light particles called **electromagnetic quanta** or **photons**. The difference between the quanta of light and of radio energy is solely that each quantum of light carries far more energy than those of radio, but in other ways they are identical.

The term 'quantum' (plural 'quanta') is a general one for any particle of energy. We can, for example, have quanta of gravitational energy (which are called gravitons) or of acoustic energy (phonons). When the energy is electromagnetic, that is involving electrical and magnetic forces, the quantum is called a photon. In what follows the terms 'quanta' and 'photons' will be used interchangeably, since this book is concerned with electromagnetics. However, because these particles are very small indeed they do not obey the laws of classical mechanics (Newton's laws), as do snooker balls, for example. Instead they behave in accordance with the laws of **quantum mechanics**, as do all very small things. This gives them some strange properties, quite unfamiliar to us from everyday life, which may even seem contrary to common sense. Two properties are important.

The first is that radio quanta can exist only when they are in motion, travelling at their one and only natural speed, which is the velocity of light. It is at present believed that nothing travels faster than this, because it is known that for anything that did time would go backwards, which seems implausible. In free space the velocity of light, always represented by the symbol c, is 299.792 456 2 million m/s, but 300 million (or 3×10^8) m/s is a very good approximation for all but the most exacting situations, and will be used in the remainder of this book.

This is the free-space value of c, but in matter (solids, liquids or gases) the speed is lower, its actual value depending on just what the matter is. In matter there is also the risk that radio quanta will collide with the atoms or molecules and give up their energy, so that as radio (or light) energy passes through matter, some energy is lost. Media range from transparent, where there is almost no loss, to opaque, where the loss is total. Again it depends on the nature of the matter concerned, but also, in a complicated way we shall look at later, on the energy of the quanta.

The second of these strange properties to take note of is that particles as small as radio quanta also have some curiously wave-like properties. (For this reason some people do not like to call them particles at all, but use made-up names like 'wavicles' or simply insist on calling them quanta and nothing else.) In fact there is no great mystery here; all things which are small enough have very noticeable wave-like properties, even particles of matter. For example, this is true of electrons, and their behaviour has to be described by means of the famous Schrödinger wave equation. However, as things get bigger their wave-like properties get progressively less perceptible, which is why we do not notice them in ordinary life.

Historically, there was a great debate between those who believed, like Isaac Newton (1642–1727), that light energy (and therefore, later, radio) was a stream of particles, and those, following Christiaan Huygens (1629–95), who thought that it was waves. Now we know that it is composed of particles (photons) which have wave properties, so there was something to be said for both points of view. Some people (and textbooks) still speak of 'radio waves', but strictly this is out-dated physics; really there are only radio quanta (photons).

In nineteenth-century France the conflict between the particle and wave theories became political, with the Left supporting waves and the Right backing particles! The mathematician Simeon Poisson (1781–1840) thought the wave theory absurd. A conservative by temperament, he was President of the Academy of Sciences and a relative of Louis XV's mistress Madame de Pompadour (born Mlle Poisson). Augustin Fresnel (1788–1827) was the leader of the left-leaning 'wave' party, and the debate between the two became acrimonious. However, the observation of diffraction effects (see Chapter 12) by an enthusiastic supporter of the wave theory, Jean Arago (1786–1853), seemed finally to prove Poisson wrong. Arago had a lively political career as a left-wing member of the Chamber of Deputies, where he advocated such radical notions as press freedom and the application of science to industry! We now think neither faction wholly right or wrong about light.

The modern understanding of radio as quantized electromagnetic energy came only in the early twentieth century, but a 'classical' theory of electromagnetics was developed in the nineteenth century by James Clerk Maxwell (1831–79), a Scot and one of the greatest theoretical physicists of all time. Although his ideas were based on defective physics, the theory that resulted is a very good approximation in most ordinary circumstances and is therefore still universally used.

The first person to observe a connection between electricity and magnetism was Hans Christian Oërsted (1777–1851) who in 1820 found that a magnetized compass needle moved when an electric current flowed in a wire close to it. The effect was studied experimentally by André Marie Ampère (1775–1836) in France, Joseph Henry (1797–1878) in the USA and Michael Faraday (1791–1867) in England. Faraday obtained detailed experimental evidence for the ways in which magnetic fields and electric currents could interact. However, great experimentalists are rarely good theoreticians so fully developing the theory proved beyond him, and in the end the task fell to Maxwell.

A Cambridge mathematics graduate, Maxwell was appointed (1856) professor at Marischal College in Aberdeen. Three years later he was made redundant, while another professor (now quite forgotten) was kept on because he had a family to support. Maxwell moved to professorial posts at King's College, London (1860) and after that Cambridge (1871). His first major scientific achievement was to formulate the kinetic theory of gases (1866), and his work on electromagnetics followed, leading to a powerful mathematical formulation of Michael Faraday's ideas about electricity and magnetism. Between 1864 and 1873 he was able to demonstrate that relatively simple mathematical equations could fully describe electric and magnetic fields and their interaction. These famous equations first appeared in his book *Electricity and Magnetism* published in 1873.

1.2 Maxwell's classical electromagnetic theory

Being uncomfortable about the notion of forces somehow acting on things situated at a distance, with nothing in between to communicate it, Maxwell chose to look at electromagnetic phenomena as manifestations of stresses and strains in a continuous elastic medium (later called the electromagnetic ether) that we are quite unaware of, yet which fills all the space in the universe. Using this idea, Maxwell was able to develop an essentially mechanical model of all the effects Faraday had observed so carefully (Torrance, 1982). His picture had the disadvantage that along with physically real things like E (electric field in volts/metre) and H (magnetomotive force in amp turns/metre) it also uses concepts like B (flux) and D (displacement) which have no real physical existence. Nevertheless it worked, predicting accurately all the electromagnetic effects that could be observed in his time, and it still works in the majority of situations, of course, although as we now know it will fail where quantum effects become significant.

Maxwell presented his equations (originally in partial differential form, but now generally expressed as four vector differential equations) that describe the electromagnetic field, how it is produced by charges and currents, and how it is propagated in space and time. The electromagnetic field is described by two quantities, the electric component E and the magnetic flux B, both of which change in space and time. The equations (in modern vector notation) are:

$$\nabla \cdot D = \rho \qquad \rho \text{ is electric charge density} \tag{1.1}$$

$$\nabla \cdot B = 0 \tag{1.2}$$

$$\nabla \times E = -\partial B/\partial t \tag{1.3}$$

$$\nabla \times H = J + \partial D/\partial t \quad J \text{ is electric current density} \tag{1.4}$$

Maxwell's equations seem incomplete:

1. The left-hand side of eqn (1.1) is the distributed electric charge density, whereas eqn (1.2) has a zero in the same place. There is no distributed magnetic 'charge' density, which would imply the existence of isolated magnetic north or south poles. But so far as we know magnetic poles always come in pairs, one of each.

2. Equations (1.3) and (1.4) are similar except for the introduction of an electric current vector J which again has no counterpart in the magnetic case. As already stated, there are no magnetic free 'charges' (poles), hence there can be no magnetic currents. These two anomalies have led to an intensive, but so far fruitless, search for magnetic currents or free magnetic poles.

The electromagnetic effects observed experimentally by Faraday (and many more beside, but not quite all) can be predicted theoretically by means of these four apparently simple equations, which was a very great triumph for Maxwell. He also calculated that the speed of propagation of an electromagnetic field is the speed of light, and concluded that light is therefore an electromagnetic phenomenon, although visible light forms only a small part of the entire spectrum.

After Maxwell's early death, Albert Michelson and Edward Morley devised experiments (1881, 1887) which showed that the ether Maxwell had assumed in fact does not exist, thus demolishing the basis of his theories. However, although the physical ideas Maxwell used to arrive at his equations were quite wrong, the equations remained a good fit to observations (in all but a very few cases). They continued to give the right answers, even though the path to them was discredited, and they remain very widely used to this day. Many textbooks avoid mentioning their inadequate physical foundations.

The principal practical problem with Maxwell's equations, however, is not their shaky physical basis, but the sheer difficulty of the mathematics that results from trying to use them: they are incapable of analytical solution in most situations of practical interest, unless it is possible to make some drastically simplifying assumptions. The alternative (more soundly based) quantum mechanical approach is usually even more intractable, however. So the rule is to use Maxwell's equations wherever you can, and quantum mechanics only where you must. Even so, because Maxwell's equations rarely lead to easy mathematics, in the past very major simplifying assumptions often had to be made to achieve acceptable analytical solutions, and this was hardly satisfactory. With the progressive fall in computing costs, this is no longer the problem it was, because solutions can be obtained using numerical methods, particularly the finite element technique. Most people who use Maxwell's equations to solve actual electromagnetic problems consequently adopt a numerical rather than an analytical approach.

In the past, textbooks on antennas devoted considerable space to analytical investigation of their properties using Maxwell's equations. In practice only very approximate solutions were possible, but it was thought necessary to demonstrate the technique and particularly some of the tricks adopted to reach solutions, which the reader might then be able to apply to other situations. A generation ago such problems provided favourite examination questions! With the advance of numerical methods our perspective has changed, and it no longer seems possible to justify finding space for any but the simplest analytical solutions. Anybody interested in the analytical approach will find that many excellent books on the subject are readily available (Kraus, 1992).

1.3 A solution of Maxwell's equations: the propagating wave

Despite all this discouragement, there do exist just a few useful analytical solutions to Maxwell's equations, and one of the most important (Fig. 1.1) is a plane wave travelling in the direction of the x-axis.

If one examines a narrow region of space (fixed x) while the wave transverses it, the electric component oscillates in strength with the **period** T (unit: seconds). The parameter f (unit: hertz), equal to $1/T$, is called the **frequency** of the wave and corresponds to the number of cycles (from maximum to minimum and back again) observed at a fixed point in one second. Examining the entire wave at any given instant (fixed time) reveals that the wave oscillates sinusoidally in space with the period λ (unit: metres). The distance λ is known as the **wavelength**. Note that the product $f \cdot \lambda$ (cycles/second multiplied by metres/cycle) must be the **velocity** of the wave (metres/second).

Accompanying the electric component is a magnetic component. The oscillating magnetic component H is perpendicular to both the electric field component E and the direction of propagation. In addition, H and E are in phase; that is, they both are at maximum

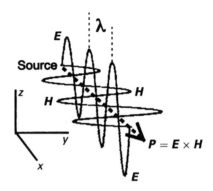

Fig. 1.1
A plane electromagnetic wave; one solution of Maxwell's equations.

amplitude at the same time. Writing $H = |\mathbf{H}|$ and $E = |\mathbf{E}|$, these two magnitudes are always proportional to each other, so that

$$\frac{E}{H} = Z_{\text{space}} \tag{1.5}$$

Z_{space} is a natural constant, equal to 120π, and having the dimensions of impedance (ohms). Some call it the **impedance of free space**, a handy way of recalling its dimensions.

The power per unit area of the wave front (the power density of the advancing wave) can be shown to be given by the Poynting vector $\mathbf{P}$ where

$$\mathbf{P} = \mathbf{E} \times \mathbf{H} \tag{1.6}$$

so

$$|\mathbf{P}| = \frac{E^2}{120\pi} \tag{1.7}$$

The magnitude of the electric field E is easily measured, so this is a useful expression.

Calculating the velocity with which the front of the wave moves forward, Maxwell found this to be c, and therefore concluded that light was just electromagnetic waves. This also leads to the most fundamental of all equations in radio:

$$c = f\lambda \tag{1.8}$$

Maxwell's conclusions, that light consists of electromagnetic waves, were in line with the scientific beliefs of his time, and seemed to have been confirmed experimentally by (among other things) the fact that the wavelength of light had been successfully measured many years before. It had been found as early as the 1820s that violet light corresponded to a wavelength of about $0.4\,\mu\text{m}$, orange–yellow to $0.6\,\mu\text{m}$ and red to $0.8\,\mu\text{m}$ ($1\,\mu\text{m} = 10^{-6}\,\text{m}$), all of which fitted perfectly with Maxwell's ideas.

It was an obvious further consequence of his theory that there might also be waves of much greater length (and correspondingly lower frequencies). Maxwell confidently predicted their existence, even though up to then they had never been observed. He died (1879) before there was experimental confirmation of this radical insight.

In 1887, Heinrich Hertz (1857–94) was the first to demonstrate the existence of 'radio waves' experimentally. He generated them by using a spark gap connected to a resonating circuit, which determined the frequency of the waves and also acted as the antenna. The receiver was a very small spark gap, also connected to a resonant circuit. The gap was observed through a microscope, so that tiny sparks could be seen.

Hertz generated radio energy of a few centimetres wavelength and was able to demonstrate that the new waves had all the characteristics previously associated exclusively with light, including reflection, diffraction, refraction and interference. He also showed that radio waves travel at the speed of light, just as Maxwell had predicted. The unit of frequency (one cycle per second) is named the **hertz** in honour of his work, cut short by his tragic death at only 37 years. He died from infection of a small wound, something which antibiotics would easily cure these days.

1.4 A quantum interpretation

We now know that the ether, assumed by Maxwell and Hertz, does not exist. There is no elastic medium for the waves to propagate in, so it follows that the waves Hertz thought he had discovered are not at all what he supposed either. What he actually generated was a stream of radio quanta, identical with the photons of light except for their energy, and small enough to have wave-like properties. Particles can perfectly well move through empty space so the ether is irrelevant to quantum theory, and it is unnecessary to make any implausible assumptions about forces acting at a distance.

Apart from its position, we can characterize the state of a particle if we specify its energy or momentum, while for a wave the corresponding parameters are frequency and wavelength. Quantum mechanics relates these pairs of parameters together, linking the wave and particle properties of quanta, in two monumentally important equations:

$$\mathcal{E} = hf \tag{1.9}$$

where h is Planck's constant and $\mathcal{E}$ is energy.

$$m = h/\lambda \tag{1.10}$$

where m is momentum.

Planck's constant, relating the wave and particle sides of the quanta, is one of the constants of nature, and has the amazingly small value 6.626×10^{-34} J/s. The tiny magnitude of this number explains why the classical theories work so well. Quanta have such very small energy (and hence mass, since $\mathcal{E} = mc^2$) and in any realistic rate of transfer of energy (power flow) they are so very numerous that in almost any situation their individual effects are lost in the crowd, and all we see is a statistically smooth average, well represented by the classical theory.

In quantum mechanics the **correspondence principle** states that valid classical results remain valid under quantum mechanical analysis (but the latter can also reveal things beyond the classical theory). However, it is good to know what is really going on (quite different from what Maxwell imagined) and there are times when thinking about what is happening to the quanta can actually help us to a better understanding.

What is the significance of the electromagnetic waves in quantum theory? From eqn (1.7) we recall that the power flow per unit area is proportional to the square of the wave amplitude

$$|P| = \frac{E^2}{120\pi}$$

But consider a parallel stream of quanta. In a time Δt they travel

$c \cdot \Delta t$ so the number emerging through a surface of area A at right angles to their flow must be

$$n\Delta t = \rho_q A c \Delta t$$

where n is the number emerging per unit of time and ρ_q is the density of radio quanta.

But each quantum carries a fixed amount of energy hf, so

$$|P| = nhf = \rho_q A c h f = \frac{E^2}{120\pi}$$

But the probability p of finding a quantum in a small volume must be proportional to the density of quanta ρ_q so

$$p \propto E^2 \tag{1.11}$$

The physical significance of the electromagnetic wave is that it tells us how likely we are to find a radio quantum, because the square of the wave amplitude (its power level) is proportional to the probability of finding a quantum near the location concerned, and the Poynting vector from eqn (1.6) just gives us the rate of flow of quanta at the point where it is measured.

When there is a flow of quanta all of the same frequency, the radiation is referred to as **monochromatic** (if it were visible light it would all be of one colour), and if it all comes from a single source, so that the quanta all start out with their wave functions in phase, the radiation is said to be **coherent**. Radio antennas produce coherent radiation, as (at a very different wavelengths) do lasers, but hot bodies produce **incoherent** radiation, experienced in radio systems as noise. By contrast, incoherent radiation is fascinating to radio astronomers, for whom hot bodies are primary sources.

In the case of coherent radiation, very large numbers of radio quanta are present, but the wave functions associated with each photon (quantum) have the same frequency and are in a fixed phase relationship, so we can treat them as simply a single electromagnetic wave, which is why Maxwell's mathematical theory works so well in practice.

1.5 The electromagnetic spectrum

Hertz confirmed Maxwell's prediction that electromagnetic energy existed not only as light but also in another form with much longer wavelengths (what we would now call radio). As a result the idea of an electromagnetic spectrum quickly developed.

For centuries people had known that the sequence of colours in the light spectrum was red, orange, yellow, green, blue then violet. By the nineteenth century this had been associated with a sequence of reducing wavelengths (or increasing frequencies) from the long-wave red to the short-wave violet. Invisible infrared waves, longer in wavelength than red, had been discovered, as also had the ultra-violet, shorter than violet. Now it was possible to imagine that electromagnetic waves might extend to much longer wavelengths than infrared. Later, when X-rays were discovered, it was also possible to fit them in as electromagnetic waves even shorter than ultraviolet. It became possible to see all the forms of electromagnetic energy as a continuous spectrum (Fig. 1.2). Quantum mechanics has not overturned this picture, but at each frequency we now add a particular value of quantum energy E.

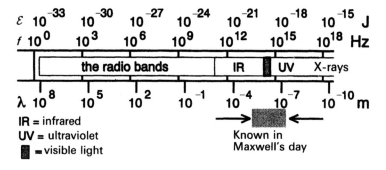

Fig. 1.2
The electromagnetic spectrum.

Table 1.1 The radio bands

Name of band	Frequencies	Wavelengths
Extra High Frequency, EHF	30–300 GHz	1 cm–1 mm
Super High Frequency, SHF	3–30 GHz	10–1 cm
Ultra High Frequency, UHF	300 MHz–3 GHz	1 m–10 cm
Very High Frequency, VHF	30–300 MHz	10–1 m
High Frequency, HF	3–30 MHz	100–10 m
Medium Frequency, MF	300 kHz–3 MHz	1 km–100 m
Low Frequency, LF	30–300 kHz	10–1 km
Very Low Frequency, VLF	3–30 kHz	100–10 km
Super Low Frequency, SLF	300 Hz–3 kHz	1000–100 km
Extra Low Frequency, ELF	30–300 Hz	10 000–1000 km

Notes: 1. In the early days of radio the LF, MF and HF bands were referred to as Long Wave (LW), Medium Wave (MW) and Short Wave (SW), respectively. These names are obsolete, but still found on mass-produced broadcast receivers. 2. Sometimes SHF is called the centimetre wave band, and EHF the millimetre wave band; together they constitute the microwave bands.

Radio technology is concerned with the lower (in frequency) part of the electromagnetic spectrum. Except at the uppermost edge of this region the quanta are insufficiently energetic to interact with the gas and water vapour molecules of the atmosphere, which is therefore transparent to radio signals. This is a great practical advantage. As a matter of convenience, the radio part of the electromagnetic spectrum is further subdivided into a series of **bands**, each covering a 10 : 1 frequency range, as in Table 1.1.

Each band has a particular range of uses and demands its own distinctive equipment designs. The mechanisms of propagation of the radio quanta in the different bands also vary enormously. In what follows, the commonly encountered means of transmission, reception and propagation will be described for all of these bands.

Problems

1. For orange–yellow light of wavelength 0.5 µm, what is the energy of the quanta? [$f = c/\lambda = 6 \times 10^{14}$ Hz. So $\mathcal{E} = hf = 4 \times 10^{-19}$ J.] (This means that a laser producing only 1 mW of light would emit 2.5×10^{15} quanta per second, and in one hour more quanta than there are stars in the known universe.)

2. A radio transmission has a frequency of 1 GHz. What is the mass of its quanta? [$\mathcal{E}$ is just over 6.6×10^{-25} J. Using the Einstein formula $\mathcal{E} = mc^2$, the mass of each quantum is $\mathcal{E}/c^2 = 7.3 \times 10^{-42}$ kg.] (The electron has a mass 10^{13} times larger. If the transmitter radiated 10 kW continuously it would take 300 years to emit 1 g of quanta.)

PART ONE

ANTENNAS

There were vast numbers of radio quanta in the universe long before Hertz performed his famous experiments. They are generated naturally whenever electric charges are accelerated or decelerated. All hot objects, in which charged particles are in rapid random motion, radiate quanta of radio energy, along with heat (infrared) quanta, and light too if they are hot enough. The stars are potent sources of electromagnetic energy, which is the basis of radio astronomy. On our own planet, atmospheric events such as lightning strikes produce showers of radio quanta, noticeable as the background crashes and crackles heard on broadcast receivers during thunder storms. In all but a very few of these cases of natural generation, the radio energy is incoherent, characterized by a jumble of quanta of very different energies. The same is true of many human-made sources of electromagnetic disturbances, such as electrical machinery and, in particular, the high-voltage spark ignition systems of petrol engines in cars and other vehicles. To terrestrial radio users, all of these just appear as noise, and their effect, if any, is harmful.

Communications engineers need to be able to launch radio quanta which, by contrast, have well-specified coherent properties, often over a very limited range of frequencies and hence quantum energies. In some cases they may wish to launch the quanta particularly in a certain direction, towards a known location where they are to be received. They do all these things by means of a structure called an **antenna**.

The first part of this book will review the principles and design of antennas.

Chapter 2

Antennas: Getting Started

Energy is supplied to the antenna as an alternating electrical current of the frequency it is desired to radiate. This alternating current is generated in a radio **transmitter** and conveyed to the antenna over a **transmission line** or **feeder** (see Appendix). An ideal antenna would radiate all the energy supplied to it, but in reality there are bound to be some losses. The radio energy supplied is partly converted into heat instead of radiated, and hence wasted. The **efficiency** of an antenna is simply the ratio of radiated power to input power, and is usually expressed as percentage. This must always be less than 100% but it can come close.

When radio transmissions are to be received, a structure is required which will intercept and absorb the quanta, converting their energy into radio frequency electrical signals which pass to a **receiver**. This too is done by means of an antenna. Like all equations in classical mechanics (from which his theories derive), Maxwell's equations remain valid if the variable t is everywhere replaced by $-t$. The same is true of quantum theory. This means that they work just as well if the direction of power flow is reversed, like a video recording played backward. Whether the direction is from power in the feeder to radiated quanta, or from quanta to power in the feeder, the same equations hold. This means that the same antenna can, in principle, be used for transmission or reception and will have broadly the same characteristics.

In reality there may be differences between transmitting antennas and their receiving counterparts, but these are of an entirely practical nature, such as the need for higher voltage insulation in transmitting antennas working at high power levels, or the need for particularly compact structures in personal radio receiving equipment. The mathematical description of the characteristics of both kinds of antennas is identical, provided they have the same configuration.

Note that 'aerial', an alternative word for antenna, is now obsolete. Note also that the plural of antenna is antennas, not 'antennae' which is a biological term.

2.1 The impossible isotrope

The simplest kind of transmitting antenna that we can conceive of is the **isotropic radiator**, or **isotrope**, which emits quanta uniformly in all directions. It can be thought of as a point in space where quanta are continuously generated (just how we shall look at later) and radiate out uniformly and equally in all directions. Conceptually, nothing could be simpler. If a sphere were to be centred on the isotrope, every unit of area would receive the same number of quanta. So, if the sphere is expressed in polar co-ordinates as being of radius r, and a small area on its surface is dA the number of quanta falling on such a small area in unit time is dN where

$$dN = \rho dA \tag{2.1}$$

where ρ is density of quanta per unit area.

For an isotropic radiator, by definition ρ is independent of both ϕ and θ, so

$$N = \int_A \rho dA = \rho \int_A dA = 4\pi r^2 \rho$$

where N is the total quanta emitted per second. Hence

$$dN = \frac{N dA}{4\pi r^2} \tag{2.2}$$

This is an important result because, as we shall see, receiving antennas capture quanta approaching them over a certain well-defined area, its value depending on the details of their structure. For a given antenna, this expression enables us to calculate the total number of quanta captured in unit time.

Sometimes it is preferable to calculate in terms of emitted and received power (rather than numbers of quanta). Since the power is equal to the number of quanta per unit of time multiplied by the energy of individual quanta (hf), the required result can be obtained simply by multiplying both sides of the equation by hf, giving

$$dP = \frac{P_T}{4\pi r^2} dA \tag{2.3}$$

where P_T is the radiated power.

This result is very widely used in calculations of radio propagation, and can be applied (with suitable modification) even to antennas which are not isotropic, as we shall see (Section 10.1).

2.2 Realising the isotropic radiator

The isotropic radiator is the simplest conceivable transmitting antenna, radiating quanta equally in all directions. The concept is useful in developing theory, but could any real antenna have this property? Obviously we shall be looking for a system with the maximum possible spherical symmetry.

Let us begin with the idea of a point (or very small sphere), isolated in space, carrying a charge q. Fortunately this is one of the cases where the classical analytical solution is easy. The electrostatic potential at range r is

$$\phi = \frac{q}{4\pi\varepsilon r}$$

where ε is the permittivity, in space ε_0.

If q varies sinusoidally with angular frequency $\omega\ (=2\pi f)$ then

$$\phi = \frac{q_0 \sin \omega(t - r/c)}{4\pi\varepsilon r} \tag{2.4}$$

where r/c is the time the field takes to reach r travelling c m/s.

This expression is known as the **retarded potential**, retarded because of the replacement of t by $(t - r/c)$. The field E at r is obtained from

$$E = -\nabla\phi$$

Since the system obviously has complete spherical symmetry, ϕ can vary only with r. So

$$E = \frac{\partial \phi}{\partial r} \cdot \frac{r}{|r|}$$

where $r/|r|$ is a unit vector in direction r, indicating the direction of E (radial).

Differentiating the above expression for ϕ with respect to r, using the formula for a product, two terms will be obtained. So we can write (the reasons for the names will appear subsequently)

$$|E| = E_r = E_{near} + E_{far}$$

where

$$E_{near} = \frac{1}{4\pi\varepsilon r^2} q_0 \sin \omega(t - r/c) \tag{2.5}$$

and

$$E_{far} = \frac{-1}{4\pi\varepsilon r}\left(\frac{\omega}{c}\right) q_0 \cos \omega(t - r/c) \tag{2.6}$$

To sum up, for this simple case of an alternating point charge we conclude:

1. That the field is radial only, and therefore the same in all angular directions from the antenna.

2. That there are two field components added together, the **near field** E_{near} and the **far field** E_{far}, where the near field varies as $1/r^2$ and the far field as $1/r$. At the same time, the far field is smaller near the sphere because of the ω/c term.

Thus, there will exist a critical value of the range such that for shorter ranges the near field will predominate, whilst at longer ranges the far field will be the larger, hence the names. The critical range will correspond to c/ω or $\lambda/2\pi$. This dimension, the **near–far transition radius**, has the greatest possible significance in antenna theory, as we shall see, but for the moment we postpone discussing it. What is quite clear is that at a few wavelengths from the source, the near field becomes quite negligible compared with the far field.

Although we have already found that an antenna which radiated equally in all directions would indeed be a useful thing, there are two major snags in trying to realize it this way. The first is simply that practically it seems impossible to build. It is easy enough to suspend a small sphere in space, but to vary the charge on it would require attaching a wire, and the charge flowing to and fro through the wire constitutes a current which would completely alter the solution of Maxwell's equations. It would certainly not be an isotropic radiator. The second snag is worse: as we have already seen, the power flow P in an electromagnetic wave is at right angles to both the E and H vectors. But in this case the E vector is radial, so the power flow therefore cannot be. Although the system is perfectly symmetrical and has a field, it does not launch electromagnetic quanta in the way we require.

In fact it is not hard to show that a truly isotropic radiating element, with radio energy flowing out from it only radially and equally in all directions, is not possible. But if a truly isotropic antenna is physically unrealisable why bother with it at all? Only because it is the most primitive antenna conceivable, with very simple properties. Even if it is impossible to build one, we can still use it as a kind of bench mark with which to compare other, more complicated, antennas. In practice people refer to it a great deal in just this way.

2.3 The isotrope as a receiving antenna

As we have seen, antennas can both transmit, emitting quanta when driven by electrical energy, or they can equally well receive, capturing radio quanta as they approach the antenna and converting their energy into electrical power which is then available to pass, through a feeder, to radio receiving equipment.

If we consider a beam of radio energy falls on an isotropic antenna, how many quanta will be captured and how many will pass right by? Each antenna is characterized by an **aperture**, or **capture area**, centred on the antenna structure. If the quanta pass within this aperture they are captured, outside it they pass by. In the case of the isotrope this boundary corresponds to the edge of the area where the near field is predominant; outside this quanta can move away freely. Near field is an induction effect, and results from the emission and almost immediate recapture of radio photons. Any radio quanta that stray within this area are very likely to be captured, whereas outside it the probability of capture falls off sharply. The radius at which the near field falls below the far field has already been shown to be c/ω, or $\lambda/(2\pi)$. Quanta that stray within this radius will be quickly reabsorbed, whilst those outside it have much less chance of being captured. It is not a sharp boundary; a few quanta will be captured from further out while a few from nearer in will escape. These two effects cancel, however, and on average it is as if all the quanta are captured within $\lambda/2\pi$ and all those beyond escape. This, therefore, is the radius of a circle corresponding to the aperture A_i of the isotropic antenna when receiving. Hence

$$A_i = \pi \left(\frac{\lambda}{2\pi}\right)^2 = \frac{\lambda^2}{4\pi} \qquad (2.7)$$

This is a very important result, because the aperture of any 'real' antenna can be compared with this basic value for the theoretical

Antennas: getting started

isotrope in order to obtain a measure of its performance. Combining it with the expression already derived for received power gives the power received by an isotropic antenna as P_r where

$$P_r = \frac{P_T}{4\pi r^2} \cdot \frac{\lambda^2}{4\pi} \tag{2.8}$$

Real antennas do better (and often very much better) than an isotrope, as we shall see. Nevertheless it is a useful standard for comparison.

Problems

1. (a) A space vehicle receiver, operating at 3 GHz with an isotropic antenna, will respond on receipt of 5×10^5 quanta. A pulsed transmitter with isotropic antenna on Earth has a peak power of 1 MW, a pulse repetition frequency of 1000/s and radiates 1 kW mean. Assuming that the vehicle cannot integrate energy between successive pulses, what is the maximum range from Earth at which it will receive signals? [8000 km]

    (b) If the thermal energy of a system is kT, what are the dimensions of the ratio kT/hf, and what significance has it [dimensionless, number of quanta needed to equal the thermal energy]? If the receiver in the first part of this question has input circuits at an equivalent of 300 K, what will be the value of this ratio, and what conclusion do you draw? [2×10^3; the input for receiver response is nearly 250 times the thermal energy] ($k = 1.38 \times 10^{-23}$).

2. A transmitter in space radiates 1 W mean at a frequency of 150 MHz, in the form of pulses with a rate of 100/s. Its antenna has isotropic characteristics. How many quanta will be received per pulse by an isotropic antenna at a distance of 1000 km? [2.6×10^{19}] What is the mean power received? [2.5×10^{-14} W]

CHAPTER 3

THE INESCAPABLE DIPOLE

The simplest practicable antenna is realized by a short straight wire, and antennas of this type are called short electric **dipoles**, or **doublets**, because they terminate in two points at which charge can collect. (Magnetic dipoles are also possible, but of these more later.) If an alternating current generator is connected into the centre of the wire dipole it can drive charge from one end to the other. What follows in this book is overwhelmingly concerned with antennas based on electric or magnetic dipoles. More complicated structures exist, such as quadrupoles, hexapoles and so on. They are hardly used at all in practice but do have interesting properties. Quadrupoles, for example, can have near field but almost no far field.

The dipole, whether short or longer, is a simple antenna that can actually be built, and it is the mainstay of radio engineering, in one form or another. Sadly, its analysis is much less straightforward than for the isotropic case. To assist understanding, the properties of the dipole will first be tackled through a traditional approach, involving the solution of Maxwell's equations.

Consider a radiating element in space in the form of a short dipole (Fig. 3.1). If it is short enough, say less than one-tenth of a wavelength, we may treat the alternating current as being of the same amplitude $I \sin \omega t$ all along the length of the dipole. Suppose it to be of length ΔL which will also be very small compared with the distance at which measurements are made (i.e. the wave is plane and radial at the measurement surface).

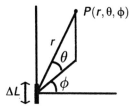

Fig. 3.1
Field of a short dipole (doublet).

To find the field at X it is necessary to solve Maxwell's equations, which is by no means easy. A short cut is to use the retarded vector potential A which is related to the current in the element by

$$A = \frac{1}{4\pi} \int_V \frac{i\left[t - \left(\frac{r}{c}\right)\right]}{r} dV$$

Since the wire is coincident with the z-axis, $|A| = A_z$ and the volume integral can be replaced by the linear integral over the range $-\Delta L/2 \leq z \leq \Delta L/2$ giving

$$A_z = -\frac{1}{4\pi} \int_{-\Delta L/2}^{+\Delta L/2} \frac{I \sin\omega\left(t - \frac{r}{c}\right)}{r} dz$$

But $\nabla . A = H$ so, assuming that $A_r = A_z \sin\theta$, $A_\theta = A_z \cos\theta$ and $A_\phi = 0$, it follows that

$$H_\phi = \frac{I \cdot \Delta L \cos\theta}{4\pi} \left[\frac{\omega}{cr} \cos\omega\left(t - \frac{r}{c}\right) + \frac{1}{r^2} \sin\omega\left(t - \frac{r}{c}\right)\right] \quad (3.1)$$

Similar expressions can be derived for E_r and E_θ and it can also be shown that all other possible field components are zero. Just as with the isotropic radiator, both a near- and a far-field component are produced, and the critical range at which the two equate is once again $\lambda/2\pi$, the near–far transition radius. However, because (still in polar co-ordinates) the magnetic field has a non-zero ϕ component and the electric field a non-zero θ component, the vector product of the two, corresponding to power flow, will have a non-zero r (radial) component. So the antenna actually does radiate radio energy. The far field corresponds to true radiation, whereas the near field is an induction effect. The far field is much the more important, but there are applications which depend on the near field also, as we shall see.

An obvious difference from an isotrope, however, is indicated by the term at the top of the expression for field, outside the bracket, which depends on cos q. This means that the dipole antenna does not radiate uniformly in all directions, in particular the radiation is zero for $\theta = +\pi/2$ and $-\pi/2$ and a maximum midway between these. This leads us to the concept of an antenna **polar diagram**.

3.1 A digression: decibel notation

To review ideas about polar diagrams, we first digress briefly to discuss **decibel notation**, which is widely used in describing antennas. This is simply a means of characterizing power ratios. If there are two powers P_1 and P_2, their decibel ratio L is defined as

$$L = 10\log_{10}(P_1/P_2) \tag{3.2}$$

in decibels (dB). (Strictly speaking a decibel is one-tenth of a **bel**, but this unit is obsolete and never used.)

The advantage of decibel notation is that it is logarithmic. Thus when two ratios are to be multiplied together their decibel values are simply added, and if they are to be divided the decibel values are subtracted. The square root of a ratio has half its decibel value, the square has twice, and so on. Some practical instances of the use of decibels are given in Table 3.1.

If ± 0.25 dB (or $\pm 6\%$) accuracy is acceptable, as it mostly is in radio calculations, it is quick and easy to estimate decibel ratios without a calculator provided the examples in Table 3.1 are memorized.

Thus, suppose we require the decibel equivalent of a power ratio of

Table 3.1 Decibel examples

Ratio	in dB
10^n	$10n$
so ...	
1	0
10	10
100	20
1000	30
1 000 000	60
	... and so on
also ...	
2	3
3	5
4 (2 × 2)	6 (3 + 3)
6 (2 × 3)	8 (3 + 5)
	... and so on
(to the nearest 0.25 dB)	

278, which is 2.78 × 100. Now 2.78 is midway between 2.5 = 10/4 and 3. But × (10/4) is (10 − 6) = 4 dB, and × 3 is 5 dB, so × 2.78 approximates to 4.5 dB, giving 278 as (4.5 + 20) = 24.5 dB (actually 24.4 dB). Bracketing the required figure with known values just above and below (and using a little judgement) estimates like this are easily formed.

The same trick works in reverse. So: 33.7 dB is (30 + 3.7) dB, and 3.7 dB is between 3 dB (× 2) and 4 dB (× 2.5), but nearer the latter. We therefore estimate 3.7 dB as × 2.3, so 33.7 dB is approximately × 2300 (actually × 2344).

If the voltage in a given circuit changes (while the circuit resistance remains the same) the change can be expressed in decibels, and often is. Because power is proportional to the square of voltage, if voltage ratios are substituted for power ratios in Table 3.1 the corresponding decibel change is doubled. Thus 10 : 1 voltage ratio is a 20 dB change (because it corresponds to a 100 : 1 power change).

However, note that the voltage ratio between two circuits of different resistance cannot be translated into decibels in this simple way; it is necessary to calculate the power in each circuit and derive the decibel figure from the power ratio, on which decibels alone are defined.

Decibels define a ratio of powers only, but they can also be used as the basis of an absolute unit of power by defining a power level as a ratio to a fixed reference level. The commonest such unit in radio engineering is the **dBm**, defined as power in decibels relative to one mW. On this basis, 1 kW may be expressed as +60 dBm, 1 W as +30 dBm, 1 mW as 0 dBm and 1 µW as −30 dBm. A much less common unit, sometimes seen, is decibels relative to one watt, written **dBw**. To convert dBw figures to dBm simply add 30, and conversely.

3.2 Antenna radiation pattern or polar diagram

For antennas generally, the density of quanta emitted in any direction is not by any means necessarily uniform, indeed that would be true only for the wholly theoretical isotropic radiator and it is certainly not so for a dipole. It is therefore necessary to be able to specify the pattern of radiation for any particular antenna (corresponding to the far-field components). Of course, this is a pattern in three dimensions, but normally it is more convenient to represent the distribution as a couple of two-dimensional diagrams, which may be in either polar or Cartesian co-ordinates. Perhaps a little confusingly, in either form these are referred to as the **polar diagrams** of the antenna, or sometimes as its **radiation patterns**.

We begin with the diagrams in polar co-ordinate form. If the antenna is at the origin of co-ordinates, one diagram represents the θ and the other the ϕ variation of the power density per unit area in the direction concerned. The radial dimension is normally power density expressed in terms of decibels relative to some convenient reference level (such as the maximum). In the case of an isotropic radiator both diagrams would, of course, be circles of

The inescapable dipole **31**

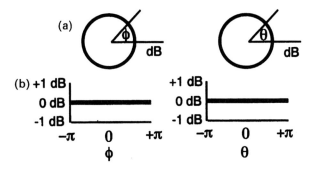

Fig. 3.2
Polar diagrams (radiation plots) for an isotropic radiator: (a) in polar co-ordinates; (b) in Cartesian co-ordinates.

constant power density (Fig. 3.2(a)). In Cartesian form these plots become simply horizontal straight lines (Fig. 3.2(b)).

For the short dipole (or doublet), assumed at the origin and vertical (as in Fig. 3.1), the ϕ diagram remains a circle, since the antenna is completely symmetrical about the vertical axis. However, for the θ plane the situation is quite different. In this plane the electrical and magnetic field components both vary as $\cos\theta$ and therefore the power (their product) as $\cos^2\theta$ (Fig. 3.3(a)). The θ-plane plot is thus characterized by zeros at $\pi/2$ and $3\pi/2$ with maxima at 0 and π. Sometimes this is called a 'figure of eight' polar diagram. As to the corresponding Cartesian plots (Fig. 3.3(b)) they are, respectively, a horizontal straight line and a raised sinusoid, though the latter looks distorted because it is plotted in logarithmic (decibel) co-ordinates.

Considering these two plots together, it is obvious that they describe a surface which in three dimensions looks a little like a ring doughnut with a very small central hole (Fig. 3.4).

This is strikingly different from the isotropic radiator, for which the counterpart diagram would be a perfect sphere. When transmitting, maximum flow of radio quanta will occur in the direction of the

Radio Antennas and Propagation

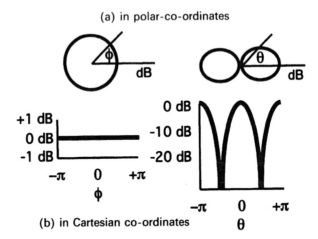

Fig. 3.3
Polar diagrams (radiation plots) for a short dipole.

median plane ($\theta = 0$ and π), which is broadside to the dipole. As θ, the angle of elevation, increases the flow of quanta gets less and less, until at $\theta = \pi/2$ and $3\pi/2$ (straight off the ends of the dipole) no quanta are emitted at all.

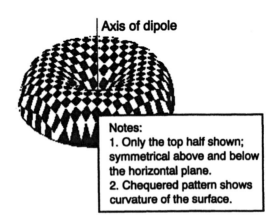

Fig. 3.4
Polar diagram of a short dipole in three dimensions.

Since, for the same power input to the antenna, the total number of quanta emitted would be the same for a dipole and an isotropic radiator, clearly the density of quanta emitted at low angles by the dipole is going to be greater than that for the isotrope, a remote receiving antenna in this direction will therefore catch more of them, and will therefore receive more signal power. This leads to the idea of an antenna having **power gain**. The name is a little misleading; no extra power magically appears from somewhere, and the antenna transmits exactly the same total power in both cases, it is just that the dipole concentrates the flow of quanta in directions at right angles to its axis, so the power density there is enhanced, and by definition the isotrope does not.

It is fairly easy to calculate how big this effect is. Suppose an isotrope radiates a total power P, then adapting from eqn (2.3), over a sphere of large radius a the power density per unit area will be p_i where

$$p_i = \frac{P}{4\pi a^2}$$

Now suppose that a dipole radiates the same total power, but that for $\theta = 0$ the power density is p_d then we have seen that

$$p(\theta) = p_d \cos^2\theta$$

As in the case of the isotrope, consider a sphere around the dipole (Fig. 3.5) and marking off the surface ring bounded by θ and $(\theta + d\theta)$.

Note that this will have an area which approximates closely to dA, where

$$dA = 2\pi a \cdot \cos\theta \cdot a \cdot d\theta$$

The total power flowing through this ring will therefore be

$$dP = 2\pi a^2 \cdot \cos\theta \cdot p(\theta) \cdot d\theta = 2\pi a^2 \cdot p_d \cdot \cos^3\theta \cdot d\theta$$

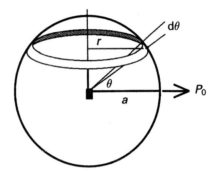

Fig. 3.5
Calculating the power gain of a short dipole.

So

$$P = \int_{-\pi/2}^{\pi/2} dP = 2\pi a^2 p_d \int_{-\pi/2}^{\pi/2} \cos^3\theta \cdot d\theta$$

The definite integral is a well-known standard form, and its value is equal to 4/3. Thus

$$p_d = \frac{3P}{8\pi a^2} = \frac{3}{2} \cdot p_i \qquad (3.3)$$

Thus in the direction of maximum radiation the dipole has a power density × 1.5 (+1.8 dB) compared with that of an isotrope. This ratio of improvement is the power gain of the antenna relative to an isotropic radiator, which we wished to calculate. This is a very modest advantage; we shall see that other antennas can have a much higher power gain. Note that this gain is attained only in the direction of maximum transmission.

The power gain of an antenna is normally quoted, as here, relative to a hypothetical isotropic antenna. Much more rarely, however, it is quoted relative to a short dipole. Since the latter has a gain of 1.8 dB, it follows that gain figures quoted relative to the dipole will be 1.8 dB less than those quoted relative to the isotrope. Often, when antenna gains are large, this difference may be too little to matter, being comparable with the margin of error in the calculations.

Because Maxwell's equations work in the same way whether the antenna is transmitting or receiving, the same power gain is also realized in the receiving mode, and more quanta approaching the antenna get converted into electrical energy in the feeder. Physically what is happening is that the dipole can capture radio quanta over a larger area (provided that they are coming from the $\theta = 0$ direction). This leads to an alternative way of looking at antenna power gain: it can be regarded as equivalent to an increase in capture area (or aperture). Since, as we have seen, the aperture of the isotropic antenna is given by

$$A_i = \frac{\lambda^2}{4\pi}$$

it follows that the aperture of the short dipole must be 1.5 times larger, and is thus

$$A_d = \frac{3\lambda^2}{8\pi} \tag{3.4}$$

a much-quoted result, but valid, of course, only for reception in the plane defined by $\theta = 0$.

In general, for any antenna we can write that if the power gain when receiving from a certain direction is G, then the aperture presented in that direction is A, where

$$A = G \cdot A_i = G \frac{\lambda^2}{4\pi} \tag{3.5}$$

3.3 Polarization

Radio quanta are always **polarized**, depending on the direction of the electric and magnetic field component of their associated wave functions. Once this polarization is established, at the time they are emitted by an antenna, they carry it unchanged until they are absorbed.

A vertical dipole will obviously give rise to an electric field vector which is vertical, while the magnetic vectors (which curl round the conductor) constitute magnetic horizontal field vectors. As a matter of convention, the emitted quanta are said to be **vertically polarized** when the electric vector is vertical. Similarly a horizontal dipole will produce a horizontal electric vector, corresponding to **horizontal polarization**. Obviously the quanta may be polarized at any intermediate angle, but this is not much seen.

Since a vertical dipole can respond only to a vertical electric field component when receiving, in principle it will capture only vertically polarized quanta, and similarly a horizontal dipole will receive only horizontally polarized quanta. In practice the separation between these two directions of polarization is not as absolute as this makes it seem; due to imperfections in their construction few antennas really have no response to the orthogonal polarization, while many quanta are absorbed by metal objects in the environment, which may then re-radiate with quite different polarisation, throwing energy into the unwanted response. It is possible to engineer for more than 20 dB separation without too much difficulty, however, and more with care.

Sometimes in transmission two differently polarized streams of quanta are combined (for example, if independently launched by two dipoles driven coherently). Suppose that one stream is horizontally and one vertically polarized, but the latter is driven by radio energy shifted in phase by 90° ($\pi/2$ radians). Since the field from the two antennas is simply added

$$\angle(E) = \tan^{-1}\left(\frac{E_y}{E_x}\right)$$

$$|E| = \sqrt{E_x^2 + E_y^2}$$

but

$$E_y = E\sin(\omega t) \text{ and } E_x = E\sin\left(\omega t + \frac{\pi}{2}\right) = E\cos(\omega t)$$

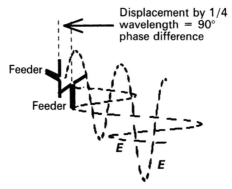

Fig. 3.6
A circularly polarized transmission results from crossed linearly polarized signals displaced in phase.

so

$$|E| = E \text{ and } \angle(E) = \omega t \tag{3.6}$$

The electric field vector is thus of constant length E and rotating with an angular velocity ω, in this case anticlockwise (Fig. 3.6). Not surprisingly this is called **circular polarization**. It could be rotating in the other direction also; the clockwise form is obtained by inverting the phase of either component. If the two components are unequal in magnitude the result is **elliptical polarization**, which is sometimes seen.

As might be expected, the receiving antenna for circular polarization can be exactly the same as the transmitting antenna, namely crossed dipoles, the output of one of which is phase shifted 90° before it is added to the other. Because the electric field can be seen as a rotating vector of constant length, it is obvious that turning the receiving antenna around its axis will not affect the received signal. This is the main advantage of circular polarization, and has led to its use on space probes and fighter aircraft, which must maintain their radio links even when rolling.

Although as with linear polarization the separation is not perfect, theoretically clockwise circularly polarized antennas will not receive anticlockwise transmissions, and conversely. This is easy to understand if we concentrate on the outputs of the two crossed dipoles. After phase shifting one through 90° they are added, but will be in phase only if the transmission is of the correct rotation, otherwise they will be in phase opposition, and will cancel.

3.4 Longer dipoles

So far we have considered only dipoles which are so short that the current in them can be treated as virtually constant. This is a good approximation if the length is up to one-tenth of a wavelength, but for dipoles longer than this it fails increasingly badly. However, longer dipoles can be expected to radiate more efficiently, and are therefore of great interest. How will they function? We could find out the hard way of course: simply solve Maxwell's equations for a longer dipole. Analytically it is rather intractable but not too difficult if done numerically, though time-consuming. Instead we will try to develop some insights into the properties of these antennas by a less punishing route, inferring solutions of Maxwell's equations without working them out directly.

In all solutions in which an ideal electrical conductor is present, like the wire of the dipole, the electric field must always be at right angles to the conductor at its surface. This is hardly surprising, since if the field had a parallel component at the surface it would cause a very large current to flow. Similarly the magnetic field is parallel to the surface when close to it. Considering the dipole (Fig. 3.7) the electric field is evidently radial to the wire, whilst the magnetic field circles around its axis.

This configuration corresponds to electromagnetic propagation along the length of the wire, a so-called 'guided-wave'. Actually it is quanta trapped near the surface that are propagating along the wire; the wave function just helps us to calculate the probability of finding them. However, we do not know whether the wave is propagating to the left or the right. Although at a point such as X the voltage must be sinusoidal, this would be just as true for a wave propagating in either direction. In the analysis that follows it will consequently be assumed that there could be propagation in both directions, and the relative magnitude of the two waves will be evaluated using a boundary condition. Thus if the generator drives a current into the right-hand half of the dipole equal to i, then

$$i = i_{L \to R} \cos \omega \left(t - \frac{x}{c} \right) + i_{R \to L} \cos \omega \left(t + \frac{x}{c} \right)$$

$$= (i_{L \to R} + i_{R \to L}) \cos \omega t \cdot \cos \left(\frac{\omega x}{c} \right)$$

$$+ (i_{L \to R} - i_{R \to L}) \sin \omega t \cdot \sin \left(\frac{\omega x}{c} \right)$$

But at the end of the dipole, where x is equal to half of l, the total current must fall to zero, so that the coefficients of both the sine and the cosine terms must equal zero. Conditions for this to be true are that

$$\cos \left(\frac{\omega l}{2c} \right) = 0 \tag{3.7}$$

and

$$i_{L \to R} - i_{R \to L} = 0 \tag{3.8}$$

From the first condition (eqn (3.6))

$$(2n+1)\pi/2 = \omega l/2c = \pi l/\lambda \quad n = 0, 1, 2 \text{ etc.}$$

so

$$l = \frac{(2n+1)\lambda}{2} \tag{3.9}$$

What this says is that the condition that the current shall fall to zero at the ends of the dipole can be met provided that its length is an odd number of half wavelengths. The shortest, and therefore the cheapest and most popular, antenna of the resonant dipole family is the **half-wave dipole**, in its various forms certainly the most widely used antenna of all.

Obviously, the half-wave dipole can only be used where its size is not excessive, and it is therefore rarely seen for wavelengths longer than 100 m; practically, therefore, its use is restricted to the HF band (3–30 MHz) and higher frequencies. At the upper end of the UHF band (300 MHz–3 GHz) 'three half-wave' dipoles are sometimes encountered, because the capture area (aperture) of a half-wave antenna becomes rather small at these wavelengths, due to its physical shortness. In what follows we will restrict consideration to half-wave dipoles, except where otherwise specified.

Because the forward (L → R) and reverse (R → L) currents must be equal at the open circuit ends of the dipole, it is a reasonable use of language to speak of the forward current being '**reflected**' at the end of the antenna. The sum of the two currents is always zero at the ends of the dipole and a maximum at the centre (Fig. 3.7). As would be expected, by a similar argument the voltage is a minimum at the

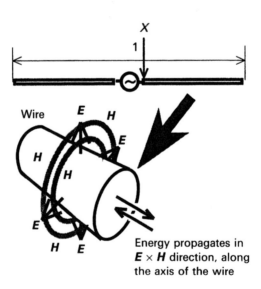

Fig. 3.7
Guided waves on a long dipole.

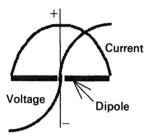

Fig. 3.8
Standing waves.

centre and a maximum at the two ends of the antenna, and in both cases the distribution follows a sinusoidal pattern, often referred to as a **standing wave**, because it is fixed in position relative to the dipole (Fig. 3.8).

3.4 Effects of changing frequency; equivalent circuits

As we have seen, for a half-wave dipole the current flowing toward the ends is totally reflected. If we consider a current starting out from the generator at the centre, it will have travelled the length of one-half of the dipole (a quarter wavelength) and back by the time it returns to the centre. This will have taken it a time equal to half a cycle of the driving voltage, so it will arrive back exactly in phase. Thus at this wavelength (and frequency $f_r = c/2l$) the antenna will appear like a resistor. At a slightly higher frequency (shorter wavelength) the half period of the driving voltage will be less but the time for the current to do the round trip will be exactly the same, since the distance and velocity are unchanged; the current will therefore lag the driving voltage and the antenna will look more inductive. Similarly at a slightly lower frequency the half period will be a little longer, the returning current will lead the driving voltage, and therefore the antenna will look capacitive.

This is all very reminiscent of resonance in an L-C-R circuit, and indeed at frequencies near those for which it is half a wavelength

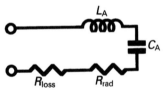

Fig. 3.9
Equivalent circuit of a dipole near resonance.

long, the dipole is well represented by a series resonant **equivalent circuit** (Fig. 3.9). It is reasonable, therefore, to refer to the frequency at which its impedance is purely resistive as the **resonant frequency** of the dipole.

Although the circuit of Fig. 3.9 is often referred to as the equivalent circuit of the dipole antenna, the two are only similar at frequencies near the resonant frequency f_r. This equivalent circuit does not, for example, predict another resonance at $3f_r$. Over a wider band of frequencies the equivalent circuit of the dipole is actually a much more complicated network, but since the antenna is primarily used only near its resonance frequencies, these simplified equivalent circuits are in order, and prove very useful.

The resonant frequency of the circuit is

$$f_r = \frac{1}{2\pi\sqrt{L_A C_A}} \tag{3.10}$$

Note that as well as the inductor and capacitor the circuit contains two resistors, which stand in for the two ways that the antenna can lose energy. One of these R_{loss}, known as the **loss resistance**, is just the usual losses we find in any electrical circuit, for example due to the conversion of energy into heat through Ohm's law effects. The second is more interesting, and is found only in antennas. The purpose of an antenna is to emit large numbers of radio quanta (photons), and each one of these carries away an amount of energy equal to hf. This is the second, and ideally the dominant, source of energy loss from the antenna, and is represented in the equivalent circuit by a second resistance R_{rad}, known as the **radiation** resistance. This can be calculated for any particular antenna configuration, and for a half-wave dipole of thin wire it is 73 Ω.

The Q-factor of the circuit is obviously

$$Q = \frac{2\pi f_r L}{R_{loss} + R_{rad}} = \frac{1}{2\pi f_r C(R_{loss} + R_{rad})} \qquad (3.11)$$

since the response of a resonant circuit falls to -3 dB relative to the peak at a bandwidth

$$\Delta f = \frac{f_r}{Q} \qquad (3.12)$$

this gives the useful bandwidth of the antenna. For wire antennas the Q-factor can be 50 (or even higher), resulting in a bandwidth no more than 2% of the centre frequency. As we shall see, however, there are ways in which this bandwidth can be extended.

Using this equivalent circuit it also becomes possible to derive the condition for maximizing the power from a transmitter which is radiated by the antenna. Thus, a transmitter can be considered as a generator, producing an open-circuit voltage e_T and having an internal resistance r_T. It is connected (Fig. 3.10) to the antenna through a series impedance X_T, the purpose of which will become

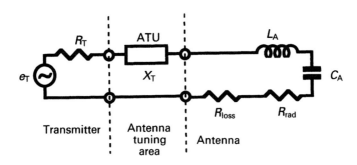

Fig. 3.10
Power matching when transmitting.

apparent subsequently. Evidently, if the current flowing in the antenna is i then the power radiated is

$$P_{rad} = i^2 R_{rad}$$

But

$$i = \frac{e_T}{r_T + r_{loss} + r_{rad} + j(\omega L - 1/\omega C + X_T)}$$

so to maximize current the first requirement is that the imaginary component of the denominator shall be zero, that is

$$\begin{aligned} X_T &= 1/\omega C - \omega L \\ &= \omega L[(\omega_r^2/\omega^2) - 1] \end{aligned} \quad (3.13)$$

where ω_r is the (angular) resonant frequency ($2\pi f_r$) given by $\omega_r = \sqrt{(1/L)C}$.

The condition for maximum current derived from the circuit reactance is therefore that there shall exist an impedance X_T, between the transmitter and the antenna, which is adjusted to a critical value, depending on the operating frequency and the antenna length, which reduces the reactance in the circuit to zero, or in other words brings it to resonance.

In the common case where the antenna is shorter than half a wavelength (the resonant length) X_T is a positive reactance, and thus may be realized by means of an inductor. By contrast, if the antenna is longer than the resonant length a capacitor is required for X_T, whilst in the commonest case of all where the antenna is cut exactly to resonant length X_T will be omitted altogether. For obvious reasons, therefore, this impedance is known as the **antenna tuning unit** (ATU), and its adjustment to remove the net reactive component is known as **tuning** the antenna. As already indicated, an ATU may not always be required, but is commonly found when, for one reason or another, it is impracticable to use an antenna long enough to resonate on its own, and is also useful if the antenna must operate over a range of frequencies. Modern ATUs use switched rather than continuously variable reactive components and are often digitally controlled, using a stored programme to set the correct values at each frequency, or alternatively using some means of detecting resonance in a feedback arrangement.

If the reactive component is cancelled out, either by cutting the antenna length to resonance or tuning the ATU, the antenna presents a resistance at its terminals

$$R_{\text{match}} = R_{\text{loss}} + R_{\text{rad}}$$

This is known as the **matching resistance** of the antenna. Note that it is always greater than the radiation resistance. What should its value be in order that the transmitter will deliver maximum power to the antenna? That power is given by

$$P_{\text{ant}} = R_{\text{match}} i^2 = R_{\text{match}} \left(\frac{e_{\text{T}}}{R_{\text{match}} + R_{\text{T}}} \right)^2$$

To maximize this value we differentiate and set the differential to zero, so

$$\frac{\mathrm{d}P_{\text{ant}}}{\mathrm{d}R_{\text{match}}} = \frac{e_{\text{T}}^2 [(R_{\text{match}} + R_{\text{T}})^2 - 2R_{\text{match}}(R_{\text{match}} + R_{\text{T}})]}{(R_{\text{match}} + R_{\text{T}})^4}$$

and, considering the numerator only, this is zero if

$$R_{\text{match}} = R_{\text{T}} \qquad (3.14)$$

which is the condition for maximum power in the antenna. This important result, that the antenna resistance must match the transmitter's for maximum power transfer, is an example of the much more general (and well-known) power matching theorem.

Sadly, not all the power delivered to the antenna is radiated, because part is dissipated in the loss resistance. The efficiency of an antenna (in per cent) is therefore

$$\varepsilon = \frac{P_{rad}}{P_{ant}} 100 = \frac{i^2 R_{rad}}{i^2 R_{match}} 100 \\ = \frac{R_{rad}}{R_{rad} + R_{loss}} 100 \qquad (3.15)$$

Antenna efficiency must always be less than 100%, since there will always be losses, but it can reach the high nineties in the case of a well-constructed resonant half-wave dipole, such as might be used in the VHF or UHF bands, and in that case the matching resistance is very little more than the radiation resistance, being typically 75 Ω. By contrast, the antenna efficiency is down to a tiny part of 1% in the ELF band, where the wavelength is so long that it is impossible to approach even a very small fraction of a wavelength with any practicable antenna. In such a case the radiation resistance is very small, and the matching resistance almost equates to the loss resistance. The overwhelming disadvantage of short antennas (that is, corresponding to very small fractions of a wavelength) is their inefficiency.

The receiving antenna (Fig. 3.11) has a similar equivalent circuit to the transmitting antenna, except that it now includes a voltage generator e_R, which represents the electrical energy derived from the captured quanta. The reactance must be tuned out, using an

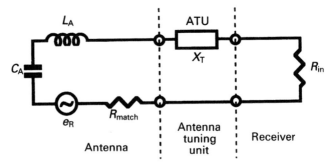

Fig. 3.11
Matching for maximum power when receiving.

ATU if necessary, and the antenna resistance matched to the input resistance of the receiver for optimum power transfer.

3.5 Polar diagram and aperture of the half-wave dipole

It is possible to calculate the polar diagram of a half-wave dipole by resolving it into an infinite number of elemental (short) dipoles and integrating (Connor 1989, Appendix E). Whereas for a short dipole we have seen that the power density $p(\theta)$ varies as $\cos^2\theta$, for the half-wave dipole the corresponding expression is

$$p(\theta) \propto \left[\frac{\cos\left(\frac{\pi}{2} \cdot \sin\theta\right)}{\cos\theta}\right]^2 \tag{3.16}$$

Superficially this looks very different; however, we note that, just like $\cos^2\theta$, it has the value 1 at $\theta = 0$ (broadside to the dipole) and is zero at $\theta = 90°$, so in these respects the half-wave dipole is like the short dipole. At 45° the short dipole is at −3 dB, while the half-wave dipole is at −4 dB. The differences between them are evidently small, and the polar diagram of the half-wave dipole is quite similar to Fig. 3.4, but with the 'ring doughnut' slightly more flattened. The radiation pattern is shown as conventional polar diagrams in Fig. 3.12.

Because power gain in an antenna depends on directing the emission of quanta in the desired direction and reducing it in other directions, it will come as no surprise that if the polar diagram of the half-wave dipole is only marginally different from that of the short dipole its power gain and aperture (capture area) will not be much different either.

Again the calculation is tedious, but the outcome is that whereas the aperture of a short dipole (from eqn (3.4) and evaluating the constants) is $0.119\lambda^2$ that of the half-wave dipole is just slightly larger at $0.130\lambda^2$. Since power gain is proportional to aperture (eqn (3.5)) it follows that the power gain of the half-wave dipole is nearly +0.4 dB relative to the short dipole. Using eqn (3.3), which

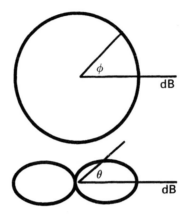

Fig. 3.12
Polar diagrams for a half-wave dipole.

indicates that the short dipole has a gain of 1.8 dB relative to isotropic, we see that the gain of the half-wave dipole is +2.2 dB relative to isotropic. The half-wave dipole does not offer so very much more power gain or aperture than the short dipole; its advantage lies in greater efficiency.

3.6 Effects of conductor diameter

So far the effects of the thickness of wire or other conductor used to construct the dipole have been ignored. We have assumed that the wire is very thin, but if this is not the case the capacitance between the two halves of the dipole is increased, modifying the values in the equivalent circuit. But from eqns (3.9) and (3.10)

$$\Delta f = f_r \cdot 2\pi f_r C_A (R_{loss} + R_{rad})$$
$$\propto C_A$$

The capacitance of the antenna is proportional to the surface area of the wires and thus to the diameter (the length being constant). Hence, approximately, Δf is proportional to the wire diameter. This increase in the effective bandwidth of the antenna is the principal effect of increasing the diameter of the wire, although there is also a small reduction in loss resistance and hence a marginal improvement in efficiency.

The velocity of propagation of a guided wave along a wire is a little slower than the velocity of quanta in free space, and the larger the diameter of the wire the more significant this effect is. As a result, a dipole is made slightly shorter than half the free-space wavelength. If the diameter of the wire is of the order of $10^{-4}\lambda$ the reduction is about 2.5%, 3.5% for $10^{-3}\lambda$, rising to 6.5% for $10^{-2}\lambda$. However, this effect is often unnoticed because of the increasing bandwidth of the antenna with increasing conductor diameter, which makes the exact length for resonance less critical.

The conductor diameter cannot be increased indefinitely, however, since the analysis of the dipole antenna has depended on the assumption that the current flow is entirely one dimensional, along the direction of the wire. If the diameter were increased too much this would no longer be true, and the antenna would begin to have quite different properties. A diameter up to one-twentieth of the length causes no significant complications, however.

3.7 Folded dipoles

Sometimes the half-wave dipole is folded, as in Fig. 3.13. The distinctive feature of a **folded dipole** is its higher radiation resistance,

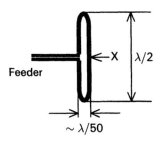

Fig. 3.13
A folded half-wave dipole.

which can sometimes make matching easier. Both sides of the folded dipole carry the same current and a moment's consideration shows that the currents are everywhere in phase between right and left. (Think of it as two close-up dipoles, identical except that one is centre-fed and the other end-fed.) It is as if it were a conventional dipole carrying twice the current. However the radiated power is proportional to the square of the current (eqn (3.9)) and is thus raised by four times, so in the equivalent circuit the radiation resistance is increased by a factor of four, bringing the matching resistance close to 300 Ω.

Another useful feature of the folded dipole is that at the point X the voltage is zero (as can be deduced by consideration of the standing wave pattern on the right-hand wire, identical with that in Fig. 3.8). The antenna can be attached to a mast or other fixing at this point without affecting its electrical properties at all, which is often convenient.

Problems

1. What is the polar diagram of an antenna? In your answer define the major features of such a diagram for polar and Cartesian co-ordinates. What is the relationship between aperture, power gain and polar diagram of an antenna? Illustrate your answer by comparing a dipole with an isotropic radiator.

2. Compare a transmitting short dipole with an isotropic radiator and show that the former has a maximum signal strength advantage of just under 2 dB at a remote point.

3. What do you understand by antenna polarization? What are the principal applications of circular and elliptical polarization, and how are circularly polarized waves launched?

4. An MF transmitter at 1.3 MHz operates into a dipole of length 100 m with a matching resistance of 70 Ω. The ATU consists of a series inductor of 100 μH. What antenna bandwidth would you expect? [37 kHz]

5. A half-wave dipole resonates at 160 MHz, has a bandwidth of 2% and a matching resistance of 75 Ω. If the operating frequency changes to 164.8 MHz, what inductor or capacitor needs to be connected in series with its terminals to bring it back to resonance? [59 pF capacitor] If the diameter of the conductors forming the dipole were halved, what would you estimate the likely bandwidth of the antenna to be? [1%]

CHAPTER 4

ANTENNA ARRAYS

Dipoles, short or long, are remarkably useful antennas, but they are limited in the range of polar diagrams that they can have, amounting to nothing beyond a more-or-less flattened 'ring doughnut' shape, as in Fig. 3.4. Because antennas achieve gain by concentrating their emission of quanta towards particular directions, this limited range of possible polar diagrams means that with simple dipoles, short or long, nothing very remarkable can be achieved in the way of power gain either.

To escape these limitations it is necessary to break out from the constraints imposed by a single dipole, and the commonest way of doing this is by using **antenna arrays**.

4.1 The simplest arrays

The simplest antenna array consists of two half-wave dipoles, shown here as vertical (Fig. 4.1) and side by side. This would be called a **two element array**. We calculate the polar diagram in the horizontal plane assuming that we are concerned only with far-field radiation and that the distance d between the dipoles is small compared with the distance r of points like P where the resulting signal is of interest. Further we assume that both dipoles are driven from the same transmitter but that the sinusoidal voltage

Antenna arrays

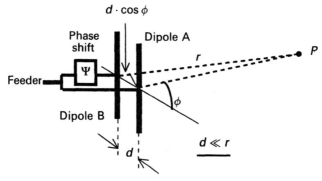

Fig. 4.1
Two dipoles as an array.

applied to dipole B is phase shifted by an angle Ψ. The field at P is

$$E_P = E_A + E_B$$
$$= k e_T \left\{ \cos \omega t + \cos \left[\omega \left(t + \frac{d \cos \phi}{c} \right) + \Psi \right] \right\}$$
$$= k e_T \left\{ \cos \omega t + \cos \left[\omega t + \left(\frac{2\pi d \cos \phi}{\lambda} + \Psi \right) \right] \right\}$$

where k is a constant (depending on r), and e_T is the voltage applied to each antenna. Using the identity

$$\cos \alpha + \cos \beta \equiv 2 \cos \left(\frac{\alpha + \beta}{2} \right) \cdot \cos \left(\frac{\alpha - \beta}{2} \right)$$

it follows that

$$E_P = 2 k e_T \cos \left(\frac{\pi d \cos \phi}{\lambda} + \frac{\Psi}{2} \right) \cdot \cos \left(\omega t + \frac{\pi d \cos \phi}{\lambda} + \frac{\Psi}{2} \right)$$

The second cosine term is just a phase-shifted sinusoidal wave function, while the first is an angle-dependent amplitude variation,

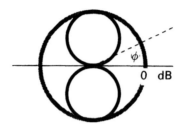

Fig. 4.2
Polar diagram of a broadside array.

which determines the polar diagram. If the power density at the point P is p, this consequently varies with the angle ϕ as

$$p \propto \cos^2\left(\frac{\pi d \cos \phi}{\lambda} + \frac{\Psi}{2}\right) \qquad (4.1)$$

By choosing suitable values of d and Ψ a wide variety of different polar diagrams can be synthesized. Two are of particular interest.

Case 1: $\Psi = 0$, $d = \lambda/2$

In this case

$$p \propto \cos^2(2\pi \cdot \cos \phi) \qquad (4.2)$$

The right-hand side $= 0$ at $\phi = 0$ and $180°$ (π radians), and $= 1$ at $90°$ ($\pi/2$ radians) and $270°$ ($3\pi/2$ radians). This is therefore a **broadside** antenna array (Fig. 4.2). An antenna like this can be useful to cover a long narrow area (for example, mounted on a bridge over a highway it will give good coverage of the road in both directions).

Case 2: $\Psi = 90°$ $(-\pi/2)$, $d = \lambda/4$

The required phase shift could be produced by an additional quarter wavelength of feeder cable, or by a suitable inductor

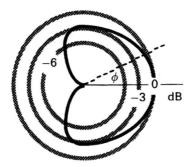

Fig. 4.3
Polar diagram of an end-fire array.

capacitor network. Now in this case, by contrast

$$p \propto \cos^2\left[\frac{\pi}{4}(\cos\phi - 1)\right] \qquad (4.3)$$

This time the RHS $= 1$ for $\phi = 0$ and falls steadily to zero at $\phi = 180°$ (π radians). The polar diagram is a **cardioid**, a heart shape, and the array has **end-fire** characteristics (Fig. 4.3).

Since the field vectors from the two dipoles simply add in the direction of maximum propagation, the received power in that direction (which is proportional to the square of the field) is increased by a factor of four (+6 dB), but since the power is shared equally between the two dipoles each receives half-power (−3 dB). Thus the overall power gain is +3 dB relative to a half-wave dipole, and thus (+3 + 2.2) or 5.2 dB relative to isotropic. Its aperture at $\phi = 0$ is evidently twice that of a half-wave dipole, or $0.260\lambda^2$.

The obvious application for this antenna array is to give a good transmission (or reception) path in the direction of its maximum, and it is therefore suited to point–point transmission or broadcast reception.

Since in both these arrays the dipoles are vertical and side by side, in the θ-plane the polar diagram is not affected for $\phi = \pm 90°$, remaining the usual 'flattened ring doughnut', although it is flattened a little more for other ϕ directions. In three dimensions, the polar diagram is evidently quite a complicated shape.

Dipole arrays like this are physically larger and heavier than single dipoles, and become unwieldy at longer wavelengths. They are therefore not used at frequencies below 3 MHz (corresponding to a wavelength of 100 m), but are compact and practicable in the upper part of the HF (3–30 MHz) band and above. If the two dipoles had been horizontal, nothing would have altered except that the θ and ϕ polar diagrams are interchanged.

4.2 Digression: talking about more general antenna polar diagrams

The notions of radiation pattern (polar diagram), power gain and aperture (capture area) which have been developed by discussing the dipole and simple arrays have a very general application to antennas of all kinds. Generally, antenna polar diagrams are very varied indeed. Purely in order to define the words that will be used to describe the commonly encountered features, consider a polar diagram of an (unspecified) antenna (Fig. 4.4), plotted in polar co-ordinates.

The **main lobe** is what the antenna is there for; it is the pattern of radiation which is used to establish a communication, measuring or radar transmission. This particular antenna has only one main lobe, but occasionally there may be two or more, where the needs of the application dictate. The width of the main lobe (sometimes called the **beam width**) is measured, in degrees, between the points at -3 dB (half power) relative to the maximum. The beam width for the simple end-fire array considered above is 180°, for example, but more complicated arrays can have a much narrower beam width.

If the beam width in one plane is $\alpha°$, whilst in the other plane the antenna is not directional, this means that in the directive plane the energy is concentrated within α rather than being spread over 360°.

Antenna arrays

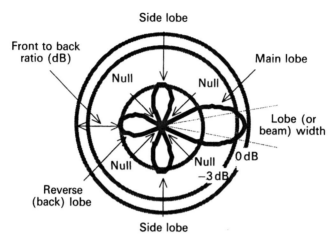

Fig. 4.4
The terminology of antenna polar diagrams.

The power gain is therefore reasonably well approximated by **Kraus's approximation** (Kraus, 1989; 1992)

$$G = \frac{360}{k_1 \alpha} \tag{4.4}$$

Here k_1 is a constant which allows for the shape of the lobe; its value is typically 1.6.

If in the orthogonal plane the antenna is no longer non-directional but has a beam width of $\beta°$, the gain is approximated by the product of the gain in each plane, so

$$G = \frac{1}{k_2} \cdot \frac{360}{\alpha} \cdot \frac{360}{\beta} = \frac{1}{k_2} \cdot \frac{1.3 \times 10^5}{\alpha \beta} \tag{4.5}$$

Once again k_2 is a shape constant, which varies from 4 for very small angles to 2.6 for large angles.

Returning to Fig. 4.4, the **side lobes** exist only with more complicated antennas; the simple dipole does not have them. Ordinarily they serve no useful purpose and the antenna designer tries to minimize them. The reverse or back lobe can be thought of as a special case of the side lobes, and it too is usually minimized. The ratio of reverse to forward radiation for an end-fire antenna, that is of main lobe to back lobe, is called the front-to-back ratio (always quoted in decibels), and is a measure of the antenna's ability to concentrate its radiation in a single desired direction.

The **nulls** correspond to angles at which no quanta are emitted at all in theory. In practice, due to minor imperfections in the construction of the antenna (such as the two dipoles not being quite parallel in the case of the simple two element arrays considered so far), a little radiation may occur at these angles, but typically this can be at a level -50 dB or even less relative to the main lobe.

4.3 A vertical collinear array

If the two vertical dipoles in the previous array examples had been one above another (**collinear** instead of side by side) (Fig. 4.5) the situation would be radically altered. In the ϕ-plane the antenna is now omnidirectional, with the usual circular polar diagram, but in the θ-plane things are more complicated.

At the point P, when $\theta = 0$ the energy received from each dipole is in phase, so the signal is a maximum. It will fall to zero, however,

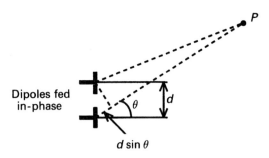

Fig. 4.5
A vertical two element array.

Antenna arrays

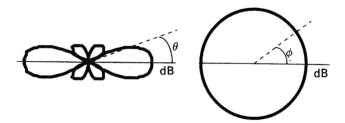

Fig. 4.6
Polar diagrams of the vertical two element array.

when $(d \sin \theta)$ equals an odd number of half wavelengths. This happens at an angle

$$\theta_{\text{null}} = \arcsin\left[\frac{(2n+1)\lambda}{2d}\right] \tag{4.6}$$

where $n = 0, 1, 2$ etc.

Since the dipoles are themselves $\lambda/2$ long, d must be greater than this. For example, if $d = \lambda$ and taking $n = 0$, θ_{null} is 30°. (Higher values of n are only meaningful with larger d since the argument of the inverse sine must be one or less.) This null is part of the θ-plane polar diagram due to the interaction of the dipoles, which has to be multiplied by the θ-plane polar diagram of the individual half-wave dipoles (eqn (3.15)). The overall result is a yet more complex polar diagram (Fig. 4.6).

The array radiates quanta in a thin disk shape all round the antenna. It is good for general coverage and broadcasting at VHF and above, putting hardly any energy into the sky, where it is not needed. Note the side lobes. The two main lobes are each about 45° (±22.5) wide, so one would expect the power gain to be near $360/(1.6 \times 2 \times 45) = 2.5$ (+4 dB). To enhance the effect, actual antennas will use more than two dipoles, perhaps six or more, in what is known as a **vertical collinear array**, and will achieve gains around 10 dB.

4.4 More complex antenna arrays

The arrays described so far, based on only two dipoles, have limited advantages and they are all **one-dimensional arrays**, that is the dipoles are arranged along a single linear axis, either vertical or horizontal. Obviously it is possible to conceive of antennas having many more active elements, and they can be arranged as a **two-dimensional array**. At the cost of increased complexity, much improved gain, aperture and directivity can be obtained in this way. Detailed analysis of the properties of more complex arrays can follow the lines used for two element arrays above; the arithmetic is naturally more extensive but not really more difficult. However, here we will simply review the properties of such arrays in general terms. As usual the two cases of interest are broadside and end-fire arrays, but the latter is the more important.

Let us begin, therefore with a two-dimensional array of end-fire antennas (Fig. 4.7), with the spacing (quarter wavelength) and relative phase (90°) between front and back dipoles needed to give this characteristic. Suppose that the array consists of m columns and n rows of dipole pairs. The polar diagram in the ϕ (horizontal) plane will depend on the number of dipole pairs in a row and their spacing, whereas the number and spacing in a column

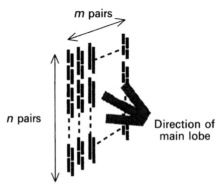

Fig. 4.7
Two-dimensional end-fire array.

sets the θ (vertical) plane polar diagram, along with the natural directivity of the dipole in this plane (the doughnut shape).

Thinking about a single row, the power gain due to horizontal-plane directivity on the line at right angles to the array plane and in front of it is equal to number of pairs, which is m, multiplied by the gain of the pair, which is 2 (relative to a dipole). Since in this plane the pairs have a cardioid polar diagram with a lobe width of 180°, using Kraus's approximation the main lobe will have a width of approximately $180/k_1 m$ degrees. In the vertical plane the gain will be n multiplied by the gain of a half-wave dipole relative to an isotrope, which, as we have already seen, is $\times 1.5$. Thus the vertical-plane gain is $1.5n$ and the main lobe width in this plane is approximately $360/(1.5n)$ degrees. The total power gain relative to isotropic is therefore

$$G = 3mn \qquad (4.7)$$

while, with suitable spacing, the horizontal and vertical main lobe widths are given, respectively, by

$$\phi_{\text{beam}} = \frac{180}{k_1 m} \text{ and } \theta_{\text{beam}} = \frac{240}{k_1 n} \qquad (4.8)$$

4.5 Adaptive arrays

So far the arrays have been assumed to have fixed phase shifts and spacing. Whilst the former cannot easily be altered the latter can, using electrically controlled phase shifters, which leads to the concept of **adaptive arrays**, antennas which have polar diagrams capable of being modified whilst in use, by means of electronic control signals. Returning to the simple two element array, from eqn (4.1) the radiated power density at a remote point is

$$p \propto \cos^2\left(\frac{\pi d \cos \phi}{\lambda} + \frac{\Psi}{2}\right)$$

Suppose that this has the value p_m at the peak of the main lobe,

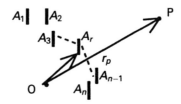

Fig. 4.8
An adaptive antenna array.

then treating the power density as a constant and differentiating

$$\frac{d\phi}{d\Psi} = \frac{\lambda}{2\pi d \cdot \sin\phi}$$

or

$$\Delta\phi = \frac{\lambda}{2\pi d \cdot \sin\phi}\Delta\Psi \tag{4.9}$$

This equation says that a change in Ψ will swing the peak of the main lobe through an angle. This is the principle behind adaptive arrays: the main lobe of an array can be swung through an angle to point it in the location where it is required. Since phase can be shifted electronically, the operation can be performed at high speed. This basic idea underlies designs for **'smart'** adaptive antenna arrays (Fig. 4.8).

We begin with transmission. Suppose there is an array of n antennas, A_1 to A_n. From some reference point O let the distance vector of the rth antenna (A_r) be r_r and let this antenna be fed with a phase shift Ψ_r. At a point P distant r_p from O the field will be the sum of that from each antenna individually, so, since $|r_p - r_r|$ is the magnitude of the distance that the wave travels from the rth antenna to P

$$E_p = E_o \sum_{\text{all } r} \cos\left(\omega t + \Psi_r + 2\pi \cdot \frac{|r_p - r_r|}{\lambda}\right) \tag{4.10}$$

where E_o is some undetermined field magnitude.

Evidently if

$$\left(\Psi_r + 2\pi \cdot \frac{|r_p - r_r|}{\lambda} = \Psi_0\right)_{\text{all } r} \quad (4.11)$$

where Ψ_0 is a constant, all the components of E_p will be in phase and therefore the received signal will be a maximum. In effect we have steered the main lobe of the antenna toward P.

From eqn (4.11) it follows that

$$\left(\Psi_r = \Psi_0 - 2\pi \cdot \frac{|r_p - r_r|}{\lambda}\right)_{\text{all } r} \quad (4.12)$$

The location of all the antennas and of P is known, so all the distance vectors are known, as also is the wavelength. Since it is only relative phases of the received field components that matter, the value of Ψ_0 can be set arbitrarily, usually so that $\Psi_1 = 0$. As a result, all of the other values from Ψ_2 to Ψ_n can be calculated, so as to place the lobe on line to P.

Antenna arrays of this kind are called **lobe-steering antennas**. They are particularly attractive because the process of lobe steering can be carried out very quickly, typically in a millisecond or less. One way in which this can be done is by using fast-operating electronic switches (Fig. 4.9). The four bit switched phase shifter shown gives $\Psi = 0$ to $180°$ in steps of $15°$. This is adequate for most arrays, but if finer resolution is wanted it only needs more control bits. The phase shift blocks can be lengths of transmission line, though at lower frequencies lumped L-C networks are more convenient.

Note that a transmission line only gives a fixed time delay (equal to the length divided by the velocity of propagation), which translates into a phase shift proportionate to frequency. Thus changing the frequency slightly will also change all the phase shifts, which gives a simple alternative way of swinging the beam, occasionally used in some lobe-steering arrays to avoid the need for changing the phase shifts by switching. However, since all phase shifts are scaled by the same factor, the condition of eqn (4.12) is only approximated

64 Radio Antennas and Propagation

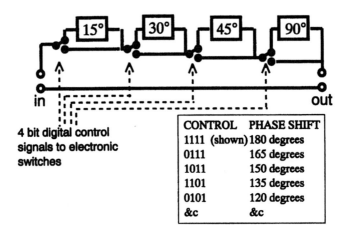

Fig. 4.9
A fast switched phase shifter.

provided that the phase changes are small. Proper electronically controlled phase shifters therefore give a more flexible design capable of superior performance, although they cost more.

The type of adaptive array considered so far is capable of steering the lobe in one plane only. This is very useful in many applications, for example for intercontinental HF radio transmission, where all that is required is to launch the signal on the right compass bearing for the geographical area where the receiving station is known to be. However, many other uses require a 'pencil' beam, covering a relatively small solid angle, to be aimed at will in three dimensions, a 3-D array. Examples of this kind of use are communication transmissions to aircraft and space vehicles, and also for radar. A two-dimensional adaptive array is capable of meeting this requirement (Fig. 4.10). For obvious reasons, arrays of this type are also sometimes called phased arrays.

The array shown has 36 elements; much larger arrays of up to several hundred elements are frequently encountered. Each elementary antenna (shown here as a vertical dipole – it would have a reflector behind it, not shown for simplicity) has a solid state power amplifier packaged behind it, as an integrated unit. Because the

Antenna arrays **65**

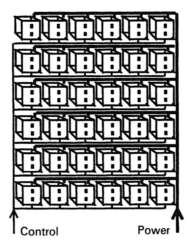

Fig. 4.10
A 36-element two-dimensional active adaptive array antenna.

amplifiers are integrated with the antenna this would be described as an **active adaptive array**. All units are identical, minimizing the cost of manufacture. Either each is driven with a suitably phased radio frequency input, or all the units are driven in phase but each has an integrated on-board controllable phase shifter. It is increasingly common for drive and control signals to be carried to the units in optical form (over optical fibres) as this avoids possible problems resulting from stray coupling between control lines and the output RF of the antenna. Each unit also receives its own power supply.

As well as making possible lobe steering, the use of a large number of small amplifiers to generate the radiated RF power in this way has another advantage. Even if a few of the units fail, the whole array still remains functional with marginal reduction in performance, so reliability is greatly improved compared with a single high-power amplifier – important for safety-critical applications such as air traffic control radar.

A two-dimensional array of this kind has a main lobe steerable both in azimuth and elevation. However, if the lobe is steered nearly

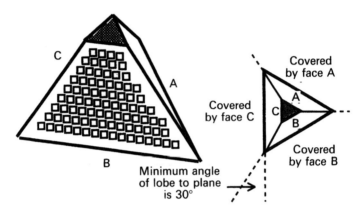

Fig. 4.11
Three or four adaptive array planes in pyramid form give omnidirectional coverage.

parallel to the plane of the array the geometry becomes unfavourable and the lobe steering is less effective (in eqn (4.9), the sine term in the RHS denominator tends to unity), so to give universal coverage, for example in radar systems, three or four such adaptive array planes are used (Fig. 4.11), in pyramid form if overhead coverage is required.

This type of adaptive array antenna can have more than one main lobe. By segregating the elementary antennas of each array face into two groups and controlling the phases in each group independently of the other, it is possible to generate two distinct lobes, for example. Each will have only half the total transmitted power and because the number of antenna elements for each is reduced the array gain for each will be reduced and the lobe will be broader. Considerations like this set a limit to the number of independent lobes which can be generated, but problems are eased by increasing the number of active elements in the array.

4.6 Adaptive arrays for reception

Receiving antennas are also made as adaptive arrays. The type of lobe-steering array already described will maximize the received

Antenna arrays **67**

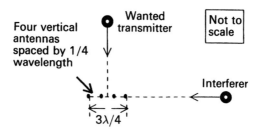

Fig. 4.12
Four antennas arranged to orient the main lobe towards the wanted transmitter and a null towards an interferer. (Seen in plan view.)

wanted signal if the main lobe is directed towards the transmitting site (and in this case there is little use for multiple lobes). However, in a receiving mode, lobe steering alone may not give optimum results. The commonest problem in radio reception is interference from unwanted transmissions. These are mostly accidental, due to the congestion of the radio spectrum, but in military use they may be intentional if the enemy uses jamming to attack transmissions. In either case, it is entirely possible for the interfering transmission to be much more powerful than the wanted signal (for example, if the interferer is nearby). To maximize the signal-to-interference ratio, it would be desirable to present a null in the antenna polar diagram towards the source of interference.

In cases where the location of the interferer is fixed it is possible to work out relatively simple antenna dispositions which will give the wanted result (Fig. 4.12). In this case the outputs from the four antennas (1 to 4) are combined in phase before being fed to the receiver. Since the antennas are equidistant from the wanted transmitter the received signals simply add; the wanted transmitter is thus at the peak of the main lobe. By contrast, the signals from the interferer reach antenna 3 half a wavelength later than those reaching antenna 1, so the combined signals are in anti phase and cancel. The same is true for the interfering signals reaching antennas 2 and 4. Thus the interferer is in a null of the polar diagram of the array and maximum signal-to-interference ratio is obtained in the received signal. Similar arrangements are obvious for more or fewer antennas. If the two signals' paths are not at right angles, as here, it will not be possible to keep the wanted signals perfectly in phase, but a few decibels' reduction in the resultant is still acceptable if a null can be steered at the interferer. The design gets increasingly unfavourable as the angle between the wanted and interfering paths decreases, however, and becomes impossible when they are coincident.

More generally, the location of the likely interferers is not known, or may change, in which event simple arrays like those described above will be defeated. In that case an adaptive array will be required, capable of changing at will the angular location of nulls in the polar diagram. Such an antenna is called a **null-steering array**, although it is invariably designed to be capable of lobe steering as well. The objective is to locate the main lobe on or near the wanted signal and the nulls on the interferers. In general, an array such as that of Fig. 4.8 will be used, but with an important difference. Since to steer the main lobe to the required direction the condition in eqn (4.12) sets the phase shift for each antenna in the array, the phase angles cannot be varied to steer the nulls, and new variables must be introduced for that purpose. An electronically controlled attenuator in each feed from the antenna to the receiver combiner makes it possible to vary the amplitude of each received signal, giving the required control (Fig. 4.13).

In this case as well as meeting the conditions on phasing given by eqn (4.12), which sets the main lobe direction, if an interfering transmitter is located (Fig. 4.8) at a position distant by the vector r_i from the receiver at O, then for there to be a null in the direction of the interfering transmission, adapting eqn (4.10)

$$0 = E_0 \sum_{\text{all } r} a_r \cdot \cos\left(\omega t + \Psi_r + 2\pi \frac{|r_i - r_r|}{\lambda}\right) \qquad (4.13)$$

where a_r is the amplitude of the rth component.

Since everything is known in this equation except the component amplitudes it is possible to select these to meet the requirement, provided (as always) that the direction of the interferer is not too

Antenna arrays

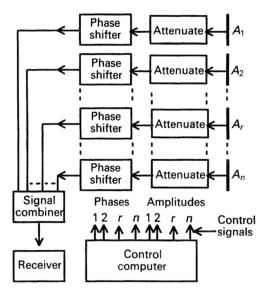

Fig. 4.13
A null-steering adaptive array having n elementary antennas.

close to that of the wanted signal. Usually one component is set arbitrarily at half its maximum value and the rest are then calculated to null this one out. Only the ratios (not the absolute values) of the components are significant in this exercise. Note that this will generally require some of the component amplitudes to be set at less than their maximum, which reduces the amplitude of their sum in the main lobe (and thus the received power from the wanted signal); however, this is accepted in view of the improvement in signal-to-interference ratio which results from nulling the interference.

Since there are $(n-1)$ independent amplitudes (one being set arbitrarily) it follows that the condition of eqn (4.13) can be met for $(n-1)$ interferers, giving $(n-1)$ nulls. However, this depends on the location of the interferers, and in unfavourable cases (when the interferer directions are close to that of the wanted signal) attempts to introduce many nulls will reduce the wanted signal magnitude.

Although the null-steering array has been described in terms of signals and interferers all in the same plane, just as with lobe-steering arrays it is possible to extend the technique to give three-dimensional coverage by using two-dimensional arrays. Modern radar antennas, for example, constructed as in Fig. 4.11, may act as lobe-steering arrays during transmission, but may also switch in controlled attenuators behind each antenna element during reception, to allow null steering. This is particularly attractive to military users, who frequently find their radars subject to enemy jamming. Even for civil radar, adaptive array antennas are now increasingly replacing mechanically rotated antennas because they are lighter, more robust and can follow rapidly moving targets. This also makes them particularly well suited for communicating with fast-moving aircraft and spacecraft.

Problems

1. Show how an array consisting of two driven dipoles can be used to produce either an end-fire or a broadside polar diagram.

2. A vertical collinear array consists of two half-wave dipoles fed in phase and spaced by half a wavelength between the pair. Sketch the polar diagram in the vertical and horizontal planes and numerically approximate the main beam width (45°). Estimate the antenna gain relative to an isotropic radiator (4 dB).

3. An array has three columns of four dipole pairs. Estimate its gain and main lobe width. (Gain = × 36 or + 16 dB, lobe width = 36°, vertical and horizontal.)

4. (a) What is an adaptive array antenna? In your answer distinguish between lobe and null steering.

   (b) An adaptive array receiving antenna operating at 3 MHz consists of four vertical dipoles in a square, all with identical apertures and spaced by a quarter wavelength.

On the axis of the array and to the right is an interfering transmitter, whilst the wanted transmitter is on a line at right angles to the array. Both are distant. Taking the top right most antenna as the phase reference, indicate the relative phasing of the other three antennas to give the best possible ratio of wanted to unwanted signals. [0, −90, 0, +90 clockwise]

5. Show why the number of distinct nulls in a plane which can be placed using a null-steering array is at most one less than the number of antenna elements.

CHAPTER 5

PARASITIC ARRAYS

In the antenna arrays considered so far all of the dipoles are connected to feeders and hence to the transmitter (or receiver, as the case may be). However, this is not essential, since it is possible to build arrays with **parasitic** elements, dipoles which are not driven directly but in which currents are induced by the near field of the driven elements. Antennas of this kind are often called **Yagi arrays**, after their inventor.

5.1 Two element parasitic arrays

Again, we begin with the simplest case of a two element array (Fig. 5.1).

The parasitic element, a continuous wire or rod cut to approximately half a wavelength, is in the near field of the driven element, a half-wave dipole. If the conductors are thin enough any coupling

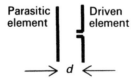

Fig. 5.1
A basic parasitic array.

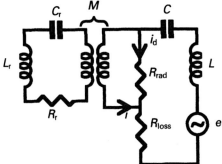

Fig. 5.2
Equivalent circuit of a driven dipole (right) inductively coupled to a reflector element.

through the capacitance between them can be neglected and only the effects of magnetic induction need be considered. The equivalent circuit of the array near resonance is therefore as shown in Fig. 5.2. The driving voltage induced in the parasitic element is

$$e_p = M\frac{di}{dt} = \frac{M}{Z} \cdot \frac{de}{dt}$$

where Z is some impedance term, which need not be evaluated.

Assuming e is a cosine wave, at resonance

$$e_p = \frac{M}{Z} \cdot \frac{de_0 \cos \omega t}{dt}$$
$$= \frac{\omega M}{Z} e_0 \sin \omega t$$
$$= \frac{\omega M}{Z} e_0 \cos\left(\omega t - \frac{\pi}{2}\right)$$

As a result a current flows in the parasitic element which (at resonance) is

$$i_p = \frac{\omega M}{ZR_p} e_0 \cos\left(\omega t - \frac{\pi}{2}\right)$$

Fig. 5.3
Polar diagram of a two element parasitic array. (C.f. Fig. 4.3.)

but the driven element current

$$i_d = \frac{1}{R} e_0 \cos(\omega t)$$

So although of slightly different amplitude, the parasitic current i_p is identical in form with the driven element current i_d except for a phase shift of $\pi/2$. Supposing that the distance between the elements d were made a quarter of a wavelength, the spacing and phasing would be identical with that in Section 4.1 Case 2, earlier (the driven end-fire array). We would expect this parasitic array to have end-fire characteristics and so it has; however, they are inferior to those of the driven array because the currents in the two elements are not equal. In the reverse direction the fields from the two elements do not exactly cancel, and a true null is consequently not found (Fig. 5.3).

In an attempt to improve the performance, the parasitic element (called a **reflector** when, as here, its function is to reduce back radiation) is moved closer to the driven element, thus increasing the mutual inductance M. However, in this case the phasing is no longer correct to give rear cancellation unless the reflector element is made inductive at the operating frequency, which requires it to be slightly longer than the resonant length.

The parasitic element absorbs energy from the driven element, part of which is wasted but most is re-radiated, and this removal of energy can be represented in the equivalent circuit of the driven element by a resistor in parallel with the usual dipole radiation resistance (Fig. 5.4).

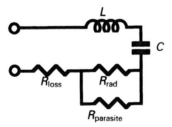

Fig. 5.4
Equivalent circuit near resonance of the driven element.

The closer the two elements are the more energy is absorbed by the parasite and the lower is $R_{parasite}$. (Note that $R_{parasite}$ is not the same as R_r; the value of M also comes into it.) There is some small loss of antenna efficiency, but in the UHF and VHF bands where antennas of this kind are used efficiency starts very high so this is not significant. More important is the change in matching resistance. At resonance, the closer the two conductors the more power the parasitic element draws from the driven element, and therefore the lower both $R_{parasite}$ and R_{match} (Fig. 5.5).

Although the polar diagram of the array is improved as d is reduced, practically speaking it is very inconvenient to have such low values of matching resistance. Most receivers or transmitters

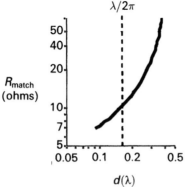

Fig. 5.5
As the reflector comes closer matching resistance falls.

are designed for a 50 or 75 Ω antenna. A commonly adopted solution is to replace the driven element by a folded dipole, described in Section 3.7 earlier. By this simple means the impedance of the driven element is immediately increased by a factor of four. The alternative of using a transformer for impedance matching is clearly less attractive.

Thus, if a two element array is required to match 50 Ω, a spacing of 0.15λ between the folded dipole and the reflector can be used (although this spacing would have corresponded to only 12.5 Ω matching resistance in the non-folded case). Note that 0.15λ is very close to $\lambda/2\pi$, the radius beyond which (as we have already seen) the field is no longer predominantly due to induction, and therefore beyond which a parasitic element will be increasingly ineffective. However, using 0.15λ will give a main lobe gain of over 5.5 dB, an aperture of $0.3\lambda^2$, and a front-to-back ratio of around 10 dB. Acceptable for less critical applications, this is a low-cost antenna.

5.2 An alternative design

As compared with a driven end-fire two element array, the antenna described above uses the parasitic element (the reflector) in place of the rear driven element. An alternative is to replace the front element, leaving the back driven. A parasitic element of this kind located in front of the driven element is called a **director**.

The analysis and equivalent circuits are closely similar to those for a reflector. Exactly the same compromise exists between having the parasitic element close to the driven element, to get good coupling, and the need to get phasing right. However, whereas moving the reflector towards the driven element was shown to be moving it in the direction of propagation and therefore had to be countered by introducing a phase lag, bringing the director closer to the radiating dipole is moving it in the direction opposite to propagation and must be countered by a phase advance. Thus, while reflectors are cut longer than the resonant length, directors are cut shorter; it is thus simply the length of the element relative to its resonant value which determines whether it acts as a reflector or a director.

Detailed analysis of the two kinds of two element arrays, one type using a reflector and the other a director, does show up some differences. The use of a director results in higher gain and front-to-back ratio for very close spacing to the radiating element, say less than 0.13λ, whilst a reflector is at an advantage for larger spacings. At the commonly used 0.15λ spacing any difference is quite marginal, and the advantage of the use of a director at closer spacings (which is anyway not large) is of little practical significance because the matching resistance is too low to be useful, even if the driven element is a folded dipole.

If an antenna is required with performance significantly better than the 5.5 dB main lobe gain and 10 dB front-to-back ratio which is easily obtained using a two element array, the solution is to move to three or more elements.

5.3 More complex parasitic arrays

By mounting both a reflector behind the driven element and a director in front of it, much improved antenna performance can be obtained at very little extra cost. Whilst still offering practicable matching resistance values using a folded dipole as the driven element. Such a three element Yagi array can better 8.5 dB of main lobe gain and an aperture of $0.58\lambda^2$, with a front-to-back ratio of 18 dB or more, using 0.15λ spacing for both elements.

The process can be taken further. The field behind the reflector is so low that there is no point in placing a second reflector there, but two or more directors, in line in front of the driven element, can give still further improvements in both forward gain and front-to-back ratio. Yagi parasitic arrays with up to six directors are commonplace, and particularly so at UHF, where they can be seen everywhere as a most popular type of television receiving antenna. As an example of how far this can be taken, a 15 element UHF Yagi array has been described with a main lobe gain of just under 16 dB, a front-to-back ratio of 22 dB and a main lobe width of about 18°. This is near the limit of what can be attained with a parasitic array using a reflector and multiple directors, because as the directors get more and more remote from the driven element the current induced in them is less, so the advantage in adding another gets progressively smaller.

Although it is solely vertically polarized antennas which have been described in this section, horizontally polarized Yagi antennas and arrays are no different in construction or characteristics (except that ϕ and θ are interchanged in the polar diagrams), and it is also easy to produce circularly polarized versions, using crossed Yagi antennas driven with suitable phases.

If still higher performance is required, the solution may be to use a number of Yagi antennas themselves arranged in a one- or two-dimensional array. This is sometimes called a **stacked Yagi array**. For example a three-by-two array of three element Yagi antennas (Fig. 5.6), if all were driven in phase, would give a gain of some 14 dB for a lobe width around 20°.

As a general rule, the more simple Yagi antennas form the array in a certain direction the narrower the main lobe in the same plane. When using vertical elements (for vertically polarized transmissions) it is commonplace to have an array with more width than height, since the basic dipole polar diagram (the 'doughnut') already contributes some reduction of vertical lobe width.

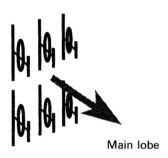
Main lobe

Fig. 5.6
A three-by-two array of 'stacked' Yagi antennas, using folded dipole driven elements.

5.4 Driven and parasitic arrays compared

Evidently, parasitic antennas can yield polar diagrams, gains and apertures only a little worse than driven arrays, whilst being cheaper to build because of the simple feed requirements. They are also lighter and offer less wind resistance, so they can be mounted on cheaper masts. Their limitation is disproportionate complexity when engineered for gains above 12 dB, but within this limit they are an attractive antenna type in the HF, VHF and UHF bands. At MF they are too large to be practicable, while at SHF and above antenna elements are so small that they become fragile, and the aperture is inadequate. An important disadvantage of parasitic antenna arrays is that the amplitudes and phase relationships in the various elements are determined by the construction, specifically element lengths and spacings. Thus, once built their properties are fixed, so neither lobe steering nor null steering is possible, other than by physically swinging the whole array.

By contrast, through variation of the phase relationships and if necessary also the amplitudes of the feeds from the various elements, a driven array can have polar diagrams varying very widely indeed, and can orient both lobes and nulls in almost any desired direction. It follows that driven arrays are at their greatest strength when one or both ends of the radio path are moving, as in radar, or where the link must be defended against accidental interference or intentional jamming, as in military radio. However, they are invariably more expensive to construct than Yagi antennas, which are best suited to applications, such as point-to-point links and terrestrial broadcasting to fixed locations, where interference is not a pre-eminent consideration and the propagation pattern is unlikely to change.

Problems

1. At 1.0 GHz (in the middle of the UHF band) a three element Yagi array is used to receive a signal from an isotropic radiator at 10 km. What is the received power, expressed in decibels

relative to the radiated power? Assume propagation in free space (about −104 dB).

2. Why do Yagi arrays have many directors but only one reflector?

3. Why do parasitic arrays frequently incorporate folded dipoles?

CHAPTER 6

ANTENNAS USING CONDUCTING SURFACES

Many antennas are constructed using a flat or curved conducting surface. Often such designs can be more economical to build than arrays of dipoles or parasitic elements with comparable polar diagrams. One way of explaining how they work is to see them as exploiting the **imaging** capability of conducting surfaces, and another is to talk in terms of the power of a conducting surface to **reflect** a quantum of radio energy. Needless to say, both explanations are compatible with each other, but sometimes one gives a clearer insight than the other; we shall develop both. We begin with imaging. An isolated small charge near a thin non-conducting plane produces varying electrostatic potential on the surface, and hence electric field components parallel to the surface (Fig. 6.1).

However, in the case of a conducting plane of very low resistivity, the whole plane must be at the same potential, despite the proximity of the charge. As already indicated, the solutions of Maxwell's equations in this case show the electric field everywhere at right angles to the plane, with no parallel component. This is exactly the same electric potential and hence field configuration as would be obtained if the conducting plane were replaced by a non-conducting plane with an equal and opposite charge symmetrically disposed behind it (Fig. 6.2). This charge may be described as the **image** of the charge that is physically present.

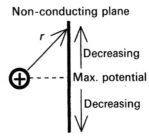

Fig. 6.1
Variation of electrostatic potential over a non-conducting surface. Potential varies as 1/r.

The thin non-conducting plane of the equivalent configuration can be considered as having no effect on electromagnetic fields, so from the point of view of electromagnetic calculations the fields can be considered determined solely by the charge and its image, as if in free space. It is not hard to show that this leads to the same solutions of Maxwell's equations as does the charge and conducting plane, and the calculations are made much easier by the substitution.

However, there are two limitations on this equivalence. First, strictly it is only valid if the conducting plane is infinite, since

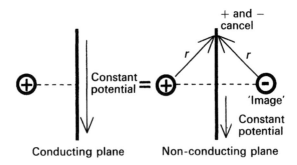

Fig. 6.2
The potential on a conducting plane is the same everywhere, equivalent to a non-conducting plane with an 'image' charge behind it, of opposite sign.

otherwise field from the charge will fringe round the edges of the plate into the region behind it. This fringing field cannot be represented by the equivalent image model. In practice it is sufficient for the plane to be large enough for the potential at its edges to have fallen to a low value. Second, the equivalent only correctly represents the field on the charge side of the conducting plane; provided that it is large enough the field behind it is negligible.

6.1 A 'plate' antenna

To demonstrate how the image concept is used, consider a half-wave dipole in front of a large conducting plane, to which it is parallel and separated by a quarter wavelength (Fig. 6.3). In practical antennas of this type, the conducting plane need not be solid, provided that the holes in it are small enough, say one-fiftieth of a wavelength or less in diameter. Often a perforated plate or metallic mesh is used, which has the advantage of less weight and lower wind forces in exposed mast-top situations.

At any instant of time an image of the charge pattern of the dipole is formed, identical with it except for the sign change, so over time

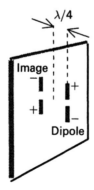

Fig. 6.3
A dipole $\lambda/4$ in front of a conducting plane creates a phase inverted image a similar distance behind it.

the currents in the image are identical in magnitude with those of the dipole. In effect, the dipole as a whole has been imaged.

We may therefore replace the dipole and conducting plane, so far as the radiation in front of the plane is concerned, by its equivalent: an array of two dipoles in antiphase, separated by half a wavelength (since the image must be as far behind the plane as the dipole is in front of it). (The plane now becomes a thin non-conducting sheet and may be ignored.) The polar diagram of such an array has already been calculated (eqn (4.1)) as

$$p \propto \cos^2\left(\frac{\pi d \cos \phi}{\lambda} + \frac{\Psi}{2}\right)$$

In this case $\Psi = \pi$ (antiphase) and $d = \lambda/2$, so

$$p \propto \cos^2\left(\frac{\pi}{2} \cos \phi + \frac{\pi}{2}\right) \qquad -\frac{\pi}{2} < \phi < +\frac{\pi}{2} \qquad (6.1)$$

Outside this angular range (behind the conducting plane) the radiation is zero if the real plane is a perfect conductor and infinite, and small even in more practical cases. This polar diagram is plotted in Fig. 6.4.

The principal advantage of an antenna of this kind, often called a

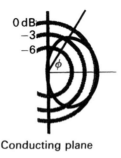

Conducting plane

Fig. 6.4
Polar diagram of the antenna of Fig. 6.3; like a twin dipole array without the reverse lobe.

Antennas using conducting surfaces **85**

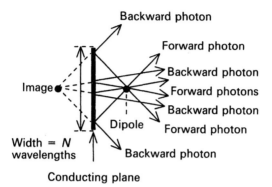

Fig. 6.5
Paths of quanta (photons). Most emitted backward are reflected and follow paths as if from the image.

plate antenna, is its high front-to-back ratio, infinite in the ideal case and large even for a more normal size of conducting plane. Whereas gain in two of the main lobes of a true two element driven array would be +5.2 dB relative to an isotropic radiator, in the case of a plate antenna one of these lobes has been suppressed, doubling the power in the remaining one to give a gain of 8.2 dB relative to isotropic, with an effective aperture of $0.52\lambda^2$. Often parasitic director elements are additionally placed in front of the dipole to improve the forward gain of the antenna and also narrow the main lobe. In this form it is widely used as a television receiving antenna.

The same result is obtained by considering the actual paths of quanta (photons) emitted by the dipole (Fig. 6.5). Quanta emitted in the forward (right) half plane are unaffected, but most of those emitted backward strike the conducting plane (all of them if it is of infinite size) and are reflected forward. From simple geometry, it is obvious that they follow the same direction and have the same path length as if they had been emitted by the image in its forward half plane. In addition, since the wave function incident on the conducting surface and that leaving must exactly cancel at the surface (to give zero probability of finding a radio quantum within the conductor), evidently the incident and reflected wave functions must be in antiphase. This may also be expressed by saying that the wave function suffers a phase inversion on reflection. Thus the quanta seem to come from the image, which must, however, be phase inverted relative to the dipole in order to give the correct phase relationship between the wave functions of quanta radiated directly in the forward half plane of the dipole and those which are reflected. So the classical and quantum descriptions of this antenna are entirely compatible, and lead to the same calculations of gain and polar diagram.

In real antennas the conducting plane will be of finite dimensions, so some quanta moving backward from the dipole will escape above and below the edges of the conducting plane, forming two side lobes. The angle subtended at the dipole by the plane, assumed to be of N wavelengths dimension, is 2β where

$$\beta = \tan^{-1}\left[\frac{(N/2)\lambda}{\lambda/4}\right] = \tan^{-1}(2N) \tag{6.2}$$

Quanta are emitted by the dipole uniformly in all directions, so the number contributing to the main lobe is proportional to the angle subtended at the dipole by the forward half plane and the conducting plate, and is equal to $\pi + 2\beta$, while the angle through which quanta escape to form each side lobe is $\pi/2 - \beta$, so the ratio of main lobe power to side lobe power is R where

$$R = 2\left(\frac{\pi + 2\beta}{\pi - 2\beta}\right) \tag{6.3}$$

This ratio (in decibels) is plotted in Fig. 6.6. In addition to side lobes, there will be some backward radiation. There are several causes, but the most important is that energy flows round the edges of the plate by a process of diffraction, discussed later. Also, if a perforated plate is used there will also be some escape of backward moving quanta through the holes. Even so, a front-to-back ratio of 20 dB is often seen.

The discussion has been developed in terms of the polar diagram in the ϕ-plane. A similar analysis applies in the θ-plane, but complicated by the intrinsic 'doughnut' polar diagram of the dipole itself,

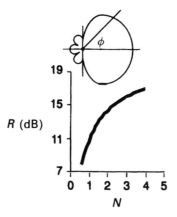

Fig. 6.6
Side lobes are R below the main lobe with a finite plate.

which has to be multiplied by the polar function generated by the plate alone.

6.2 The corner reflector

A development from the simple plate antenna is to use two conducting planes set at an angle, partly enclosing the dipole (Fig. 6.7). Indeed, it is sometimes helpful to regard the plate

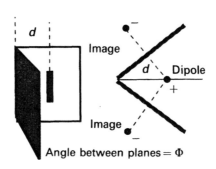

Fig. 6.7
A corner reflector.

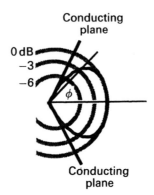

Fig. 6.8
Ideal polar diagram of a 120° corner reflector. In real life there will be small side and backward lobes.

antenna as a special case of this configuration when the included angle is 180°. This type of antenna is known as a **corner reflector**. In principle any angle between the two planes may be chosen, but 120° and 90° are by far the commonest, with 60° seen occasionally. In all cases the analysis of the antenna may use the image model or rely on consideration of the paths taken by radio quanta, with exactly the same results.

A typical polar diagram for a corner reflector, in this case with a 120° angle, is shown in Fig. 6.8. The polar diagrams of all antennas of this type have certain features in common. Taking, once again, the case where the dipole is vertical and considering the ϕ-plane polar diagram, there will always be nulls in the direction of the conducting surfaces, supposing them to be infinite, since there is no possible direction of emission of radio quanta from the dipole which will cause their path after reflection to be tangential to the plane. By contrast, the main lobe is on the line bisecting the angle between the planes, and is a maximum provided that the radio quanta reflected into the main lobe have a path length half a wavelength longer in the forward direction than those emitted directly (thus allowing for the phase inversion at reflection). In image terms (Fig. 6.7) the images must be half a wavelength further back than the dipole. From this it is easy to calculate the value of d,

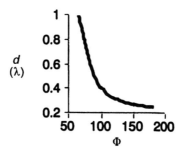

Fig. 6.9
Best vertex-dipole distance versus angle.

the displacement of the dipole from the vertex, as a function of F, the angle between the conducting planes.

The distance from the dipole to the plate is $d\sin(\Phi/2)$, so the images are $2d\sin^2(\Phi/2)$, and the condition for a main lobe maximum is (Fig. 6.9)

$$d = \frac{\lambda}{4\sin^2(\Phi/2)} \qquad (6.4)$$

Bearing in mind that the conducting planes must extend well forward of the dipole to avoid undesirable side lobes and reverse radiation, the rapid rise in size required as the included angle is decreased is a powerful reason not to choose smaller angles. Sometimes the distance d is reduced a little below the optimum value indicated above, to broaden the main lobe.

The corner reflector produces a wide main lobe. This is useful for area coverage, where the location of the other end of the radio link is not precisely known (for example, in broadcasting and mobile-phone base stations) but for many applications a much narrower lobe would be an advantage. This can be achieved by moving from flat conducting planes to curved ones.

6.3 The parabolic trough reflector

The narrowest possible main lobe from an attenuator would be achieved if all the radio quanta it emitted travelled parallel to each other, instead of being at a range of angles as with the antennas considered hitherto. To achieve the desired parallel beam of photons using conducting surfaces as reflectors, it is necessary to move from flat to curved surfaces (Fig. 6.10). What is the correct curvature of the surface and where exactly should the dipole be located to give this desirable result?

If the beam is thought of as propagating in the x-direction of Cartesian co-ordinates (Fig. 6.11), the reflecting surface will have a form

$$x = g(y)$$

What can be said of the function g? Clearly it must be symmetrical about the x-axis, so if g is a power series it can only contain even powers (mathematically, g is an even function). We begin by seeing whether the simplest function of this type, a square law, will meet

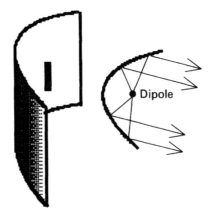

Fig. 6.10
A suitably curved conducting surface is capable of producing parallel photon paths.

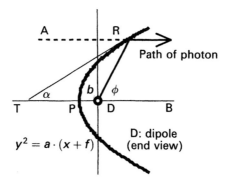

Fig. 6.11
Analysis of the reflecting surface shows it to be a parabola in section.

the requirement. A suitable expression for the curved surface might be

$$y^2 = a(x+b)$$

where a and b are constants.

We can identify certain features of the diagram using this expression. At $y = 0$, $x = -b$ so b is identifiable as the distance from the vertex P of the curve to the location D of the dipole. Also

$$\frac{dy}{dx} = \frac{a}{2y}$$

When R is directly above D, so that ϕ is a right angle and $x = 0$, TR must be at 45°, corresponding to a slope of unity, so

$$(y^2 = ab \text{ and } a = 2y)_{x=0}$$

Hence $a = 4b$, so the equations of the surface and gradient become

$$y^2 = 4b(x+b) \text{ and } \frac{dy}{dx} = \frac{2b}{y} \tag{6.5}$$

The point D, distant b from the vertex of the curve, is known as the **focus** of the reflector and the distance b (which may be obtained from eqns (6.5)) is its **focal length**. The dipole at the focus is often called the radio **feed**, and is said to **illuminate** the reflecting surface, often called a **secondary reflector**, because (unlike a reflector element in a driven or parasitic array) it is reflecting radio quanta which may already have escaped from the near-field region of the antenna.

A little geometry will now be helpful. The line TR is a tangent to the curved surface at the point R at which the radio quantum following the path DR is reflected to the horizontal, along an extension of the line AR to the right. The angle BDR (ϕ) is equal to the angle ARD, so bearing in mind that for the radio quantum being reflected the angle of incidence is equal to the angle of reflection, it follows that the tangent TR must bisect the angle ARD, so angle ART is $\phi/2$. But this is equal to angle RTP (α). Evidently, if it can be proved that $\alpha = \phi/2$ for all ϕ then eqns (6.5) describe a surface which has the required property that whatever direction photons leave the dipole they will leave the antenna, after reflection, parallel to the x-axis.

Now

$$\tan \alpha = \left(\frac{dy}{dx}\right)_{at\,R} = \frac{2b}{y}$$

However

$$\tan 2\alpha = \frac{2 \tan \alpha}{1 - \tan^2 \alpha} = \frac{2(2b/y)}{1 - (4b^2/y^2)}$$

$$= \frac{4by}{y^2 - 4b^2} = \frac{4by}{4b(x+b) - 4b^2}$$

$$= \frac{y}{x} = \tan \phi$$

This is the result we are looking for, showing that the surface defined by eqns (6.5) has the desired property of focusing the radio quanta emitted by the dipole into a parallel beam. This curve is a **parabola**.

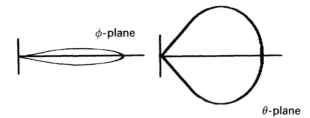

Fig. 6.12
Polar diagrams for the parabolic trough antenna.

An antenna such as that shown in Fig. 6.10, often called a **parabolic trough antenna**, has the property of focusing the radio quanta into a parallel beam in the ϕ-plane, provided that the dipole is located at the focus. The result is a ϕ-plane polar diagram with a very narrow main lobe, indeed if all of the photons travelled parallel to each other the lobe would be of zero width – the case only for an infinite reflecting surface. A real antenna of finite size will have a lobe of non-zero width. Of course, in the θ-plane there is no curvature of the reflecting surface, so the θ-plane polar diagram is like a plate antenna (Fig. 6.12). In three dimensions the polar diagram might be described as fan shaped. Although such antennas are sometimes useful, for example, in height finding radars where the beam must be narrow in the vertical (elevation) direction but can be wide in the horizontal (azimuth) axis, they are not common.

Sometimes in this and other antennas using a flat or curved conducting plane, the conducting sheet is replaced by a grill of conducting bars, all parallel to the antenna element and separated from each other by a very small fraction of a wavelength. For incident radiation cross-polarized relative to the bars very few of the radio quanta will be reflected, since a full image cannot form at right angles to them. The effect of this, particularly important when receiving, is that the gain-enhancing properties of images are present only for received waves polarized parallel to the bars. This is a useful way of enhancing the polarization discrimination of the antenna, and by using this technique 36 dB of polarization separation has been reported.

6.4 The paraboloid reflector

The parabolic trough reflector, considered above, is curved on one of its axes but flat on the other, and produces a fan-shaped main lobe. By curving the surface in both directions the polar diagrams can be made the same in both axes, corresponding to a main lobe narrow in both ϕ and θ. This is sometimes called a **pencil beam**, and is useful where the locations of both transmitter and receiver are known or can be found. Ideally the main lobe will be identical in both planes, indeed in a three-dimensional representation its cross-section will be circular.

To achieve this, the parabola shape of Fig. 6.13 is rotated around the x-axis to produce a paraboloid of revolution. This dish or bowl-like shape is called a **paraboloid reflector**; placing a small radio feed, such as a dipole, at its focus results in the desired pencil beam.

Because the feed, shown here as a dipole, itself has an asymmetrical polar diagram, the cross-section of the main lobe will be elliptical rather than circular, but the effect is small. A more serious disadvantage of a simple dipole feed is that many of the radio quanta it emits will not go to illuminate the parabolic reflector but

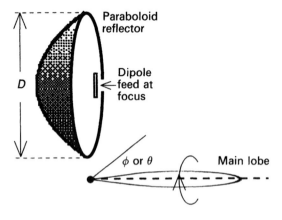

Fig. 6.13
A paraboloid reflector produces a pencil beam, the polar diagram almost symmetrical about its axis.

will miss it altogether (for example, those emitted forward). To overcome this problem, the dipole feed is replaced by a small array having a main lobe just wide enough to illuminate the reflector. This may be a Yagi array, or a plate antenna, and concentrates radio quanta onto the parabolic reflector. All (in theory) find their way into the main lobe and energy is not wasted. In practice, paraboloid reflector antennas always have small side and backward lobes, due to minor imperfections, but these can be tens of decibels down on the main lobe.

The characteristics of the paraboloid antenna are very easy to calculate. Consider it as a receiving antenna, looking into the axis of the main lobe, all radio quanta travelling parallel to the axis which are intercepted by the paraboloid will be directed to the feed antenna, where they will be captured. Thus the aperture of the antenna should be simply the area presented by the circular rim (the 'mouth') of the paraboloid. In practice there are some small fringing effects and other losses which make this slightly optimistic, but not to any significant degree. Thus the aperture of the antenna is given by

$$A_p = \frac{\pi D^2}{4} \tag{6.6}$$

Dividing this aperture by that of an isotropic radiator gives the gain of the antenna relative to isotropic, which is just

$$G_p = \frac{\pi^2 D^2}{\lambda^2} \approx 10 \left(\frac{D}{\lambda}\right)^2 \tag{6.7}$$

This expression is often misunderstood. It says that the power gain of an antenna of given size increases as wavelength gets smaller, but bear in mind that this power gain is defined relative to an isotrope, the effective aperture of which falls as the square of wavelength, so no advantage is gained in the antenna's ability to launch or capture radio quanta by reducing their wavelength. The aperture is the best guide in this respect, and eqn (6.6) shows it to be wavelength independent, as one would expect.

The angular width of the axially symmetrical main lobe (using Kraus's approximation, eqn (4.5), with the shape constant equal to four, for small angles) is

$$\phi = \sqrt{\frac{1.3 \times 10^5}{4G_p}} = \frac{3.25 \times 10^2}{\sqrt{G_p}} \approx \frac{100\lambda}{D} \quad \text{(degrees)} \quad (6.8)$$

This result serves to point up the unique characteristics of the antenna. It can have a very large aperture compared with a dipole at the same frequency, so when receiving it will capture a large number of radio quanta and hence produce a relatively big signal at the receiver input port. When transmitting, the power gain is proportionately large. However, the polar diagram of such an antenna consists principally of a main lobe of very small cross-section. If we know where the target receiver is this narrow radio beam is no disadvantage, so that, for example, it is widely used in fixed point-to-point radio links, and also for broadcasting satellite receivers.

It may also be useful even for transmitting to a receiver that is not so well located provided the transmission distance is large, so that the narrow main lobe spreads over that distance to cover an acceptable area. Examples would include communication with distant space vehicles and for transmission from distant satellites to Earth. A direct-broadcasting Earth satellite, roughly 37 000 km away, would cover a patch of the surface at least 650 km in diameter per degree of main lobe width. Given a population density of 50 persons per square kilometre (easily surpassed for most of western Europe and on the US eastern seaboard) the antenna could provide broadcasting services to a total population of over 16.5 million people per degree, and increasing as the square of the lobe width. (This assumes that the coverage area is at or near the equator. Since the satellite will be in the equatorial plane, at higher latitudes the coverage zone becomes elliptical, and its area increases still further.)

Another important application of paraboloid reflector antennas has been in radar. Here things work the other way round. The position of the target, such as an aircraft, is initially unknown, but the antenna is scanned mechanically, in both azimuth and elevation, until a signal is received from it. Because of the very narrow main lobe, the orientation of the antenna when a signal is received indicates the direction vector of the target. (Its distance is determined by, for example, the time it takes a radio pulse to go from the ground transmitter to the target, be reflected and return.) This is still a very common type of radar antenna, but is now being slowly eclipsed by the adaptive array 'smart' antennas described in Chapter 4.

The main disadvantage of the paraboloid antenna for radar applications is the need for mechanical scanning of the target area, which involves physically moving the antenna, usually by means of servo-motors and gear trains. The speed with which targets can be acquired and followed is seriously limited by the mechanical inertia of the scanning system. Also mechanically scanned antennas have only one usable lobe (whereas 'smart' antennas can have many) which limits their capacity to engage multiple targets. They do not have independently steerable nulls, so are more vulnerable to jamming and interference. They are also relatively heavy because of the mechanical components (a particular disadvantage on ships and aircraft) and are subject to increased possibility of damage by blast or heavy weather. Until now their principal advantage has been lower first cost, but the cost of active adaptive array antennas is falling, and must end up well below that of mechanically scanned paraboloids. The days of the familiar parabolic 'dish' radar antennas are therefore numbered, although they will doubtless survive for many years yet.

6.5 Unavoidable conducting surfaces: ground-plane antennas

Sometimes an antenna must be designed to work in conjunction with a conducting surface which is already present, and which will therefore have an unavoidable effect. Obvious examples are antennas mounted on the metal skins of aircraft, cars, trucks, trains and ships. Then again, in the MF band and at lower frequencies, where wavelengths range from hundreds up to many thousands of metres, it is not practicable to lift antennas many wavelengths above the Earth's surface, land or sea, so it will ordinarily be necessary to regard the Earth as a conducting plane and design the antenna system to incorporate it. Such antennas are often described as **ground-plane** or **Marconi antennas**. The term monopole is also used for them.

> Guglielmo Marconi (1874–1937) was the Italian son of Annie Jameson, an Irish woman who was the most formative influence in his life. In 1894 he repeated Hertz's radio experiments, making improvements. Two years later he moved to London and formed Marconi's Wireless Telegraph Company (1897). The first radio message across the Channel was sent by Marconi in 1899, and the first transatlantic contact, from Cornwall to Newfoundland, was on December 12th, 1901. He shared the Nobel Prize for physics in 1909. When he died Marconi was accorded the unique tribute of a worldwide two-minute radio silence.

How is the design handled in all these ground-plane antennas? In what follows we shall simplify the argument by assuming that the conducting surface is flat. In practice it need be so only for a few wavelengths around the antenna, and even modest degrees of irregularity or curvature have a small effect on the antenna characteristics.

The skin of a vehicle could certainly be used to form a plate antenna, as in Section 6.1 earlier. The disadvantage of this antenna is that it produces only a broad lobe at right angles to the vehicle, which is often not what is required. Even so this kind of antenna is sometimes seen, particularly on buses and trains, where communication is required only with track-side radio 'beacons', often used for vehicle location purposes and to pass local running instructions.

More commonly, all-round radio coverage is what is required, since the vehicle can be heading in any direction. Omnidirectional coverage is mostly obtained by using a quarter-wave or three-quarter-wave vertical antenna mounted on a convenient horizontal surface, such as a vehicle roof (Fig. 6.14). They are popular from

Antennas using conducting surfaces **99**

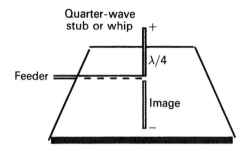

Fig. 6.14
A quarter-wave antenna perpendicular to a conducting plane, often called a stub or whip antenna.

the HF to the UHF bands, in which service they are variously termed quarter-wave **stub** or **whip** antennas. UHF versions mounted on an aircraft are often made solid, with a streamline cross-section to reduce aerodynamic drag, in which case they may be called **blade** antennas. The feeder is connected between the conducting plane and the (insulated) base of the antenna.

The usual phase-inverted image is formed in the conducting plane; it can be thought of as providing the 'missing half' of a complete half-wave dipole. The polar diagram is thus like a dipole, except that quanta are emitted solely in the region above the plane. The picture is therefore of the upper half only of a 'doughnut' shape, similar to Fig. 3.4, though this is for a short dipole and a quarter-wave whip will be slightly more flattened (Fig. 3.12, upper half only). Clearly, this type of antenna must be polarized perpendicularly to the plane, which means vertically if the plane is horizontal.

The matching resistance of a quarter-wave whip is less than the 75 Ω of a dipole. The solution is simple if we consider what is actually happening. In the transmitting case, if a certain RF voltage is applied to the antenna the radiation will be as if from the dipole formed by the antenna and its image. However, the quanta which seem to come from the image in fact come from the antenna and are reflected by the conducting plane, so the antenna must radiate twice as many quanta, which implies that the current increases by the

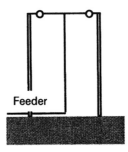

Fig. 6.15
A Marconi type T-antenna, supported by two masts.

square root of 2 relative to the true dipole case, so the resistance must be reduced by the same factor. This shows it to be near 52 Ω.

Vertical antennas directly over a conducting ground plane (electrically closely similar to whip antennas) are virtually the only option available in the MF band and at still lower frequencies, and it is almost always impracticable for them to be as long as a quarter wavelength. Many forms are in use. In his early experiments Marconi adopted a near-vertical wire held aloft by a kite, and wires sustained by kites or balloons are still occasionally used as temporary installations, but most are sustained by masts.

The top wire of the **T-antenna** (Fig. 6.15) plays no direct part in receiving the vertically polarized transmissions, needless to say. Where the vertical wire is considerably short of a quarter wavelength, its purpose is to provide a 'capacitor top' to the antenna. This increases the capacitance of the top end of the vertical wire to the ground plane, which improves the current distribution (since the current need no longer fall to zero at the top of the vertical wire, but only at the ends of the horizontal) and makes it easier to bring the antenna to resonance, by reducing the value of the series inductor needed in the antenna tuning unit.

The **mast antenna** is widely used in MF/LF broadcasting, because it can be engineered for high-power transmission (Fig. 6.16). At the bottom end of the AM broadcasting band the wavelength is 200 m,

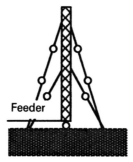

Fig. 6.16
The mast antenna stands on an insulating footing, supported by guy wires broken into sections by insulators.

and a quarter-wave mast height is possible. However, with the cost of a mast increasing roughly as the cube of its height, at longer wavelengths this gets rapidly more expensive and soon becomes prohibitive. Thus for most of the MF band as well as for the lower bands (LF and VLF) mast antennas are of less than resonant length, and are tuned using an ATU. Sometimes wire structures supported by radial arms are constructed at the head of the mast to provide a capacitor top, making the ATU design easier and improving efficiency a little. The mast itself, which is carefully bonded at all constructional joints to ensure low resistance, stands on an insulating foot, where it joins the feeder. Although sometimes free-standing, it is much more often supported by guy-wires, themselves broken by insulators into lengths short enough to ensure that dangerous voltages are not induced.

These are typical MF/LF antenna structures, although there are innumerable variants on the theme of vertical monopoles. All antennas of this kind assume that the ground can be treated as a perfectly conducting plane. At sea this is reasonable, but on land soils are very variable and may not have a very high conductivity, particularly if dry and sandy. To ensure satisfactory functioning of the antenna, and avoid dangerous and wasteful potential gradients over the ground in the case of high-power transmissions, wires may be laid over the surface to ensure that it is a true equipotential. These may be in the form of radial wires extending from the foot of a mast antenna, often buried a few centimetres below the surface; such an arrangement is called a **counterpoise**. In an alternative version, buried wire mesh may be used.

6.6 Ground-plane arrays

In principle it would be possible to duplicate all the types of dipole arrays using ground-plane antennas. Thus it would be possible to put earthed parasitic elements close to ground-plane antennas to modify their polar diagrams. MF broadcast antennas like this are sometimes built, but more often an omnidirectional response is required for this service.

The only growing use of ground-plane antenna arrays is for adaptive arrays in the HF and MF bands. By using large arrays, typically consisting of 20 or 30 whip antennas arranged equispaced around a circle many wavelengths in diameter, it is possible to produce a fully steerable high-gain main lobe whilst at the same time directing several antenna nulls at sources of interference. Computer-controlled systems of this kind, though large, complex and expensive, cannot be surpassed for intercontinental reception in these bands.

Problems

1. A plate antenna has a plate-to-dipole spacing of 0.35λ. Sketch the polar diagram and calculate the significant angles. [two lobes at $\pm 45°$, -1 dB on axis]

2. A paraboloid antenna has a diameter of 60 cm and operates at a frequency of 30 GHz. What will be its characteristics? [aperture $= 0.29$ sq m, gain $= 45$ dB, lobe width $= 1.5°$] At what frequency would a half-wave dipole have the same aperture? [200 MHz] How would its polar diagram differ? [not pencil beam but 'doughnut']

CHAPTER 7

WIDE-BAND ANTENNAS

All the antennas considered so far are operated in a resonant mode. This may be achieved by cutting the elements to a critical length or through the use of an ATU. By bringing them to a resonant condition the current flowing in the antenna is maximized for a given applied voltage, and for this reason resonant antennas are preferred wherever possible. However, they have the disadvantage that, as with all resonant circuits, the operating bandwidth is limited. As we have seen (Section 3.4) this can result in satisfactory operation over a frequency range which is no more than a very few per cent of the resonant frequency. For conventional radio transmissions with a very small fractional bandwidth this may not matter but there are other categories of use where it becomes a serious disadvantage. There are two classes of application for which wide-band antennas must be used. These are:

Where the transmission carries a very wide-band modulation. Examples include:

- Very short pulses, which have a wide Fourier spectrum and are used in some precision radar systems to give very fine spatial resolution.
- Spread spectrum and pseudo-noise modulations, such as the Rademacher–Walsh waveforms used to modulate Earth-penetrating radar, and direct sequence spread spectrum modulation, used in some communication systems.
- Barrage jamming transmitters, used by the military to

interrupt radio communication over the range of frequencies the enemy might wish to use.

And, more commonly, where the transmissions are required to change frequency often and over a substantial range. This occurs:

- When a number of different transmissions are required to be selected, spread over an extended frequency band, as in television broadcasting.
- If changing atmospheric conditions force the selection of a new operating frequency from time to time through the day and depending on the seasons and other factors. This is typical of ionospheric propagation in the HF band.
- In the case of frequency hopping transmissions, which change their carrier frequency many times per second in a pre-programmed way.

For all of these applications there is considerable interest in broadband antennas which retain reasonable efficiency. Many different approaches can be taken to this problem.

7.1 What can be done with dipoles and monopoles

We have already noted that the bandwidth of a dipole antenna, rarely 10%, can be extended by building it with conductors of larger diameter. The same is true of monopoles. It is possible, for example, to build a quarter-wave stub in the form of an inverted cone (Fig. 7.1) to give a wide bandwidth. However, there are severe practical limits to this approach, of which cost, weight and size are not the least important, so it tends to be limited to shorter wavelengths, where the resulting antenna is physically not too large, say in the VHF band (30–300 MHz or 1–10 m wavelength) and at higher frequencies.

It is also possible to operate with dipoles much shorter than their resonant length, bringing them to resonance by using an ATU. Adjusting the ATU can be used to obtain resonance at different frequencies, thus covering a band, and this is conveniently done

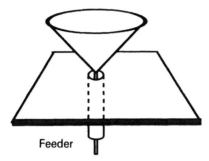

Fig. 7.1
A wide-band conical Marconi antenna. A bi-conical dipole version is also sometimes seen.

under microprocessor control. Similar solutions can be applied to monopole antennas, which (as already noted) operate satisfactorily at sub-resonant length using an ATU. This approach applied to whip antennas on military vehicles, for example, makes it possible to obtain satisfactory HF working over a band of frequencies. The snag is that short antennas have much lower efficiency than their resonant counterparts and at any given tuning point have even narrower bandwidth, because their low radiation resistance results in a high Q-factor.

If the changes in operating frequency can be relatively slow and coverage of the band at just a few spot frequencies is acceptable, it becomes possible to use resonant dipoles cut for several different frequencies and switch between them. If the resonance at three half wavelengths is exploited as well as the fundamental resonance, a band can be reasonably well covered with just a few switched antennas. To cover spot frequencies across the band 5–30 MHz (upper HF band) antennas cut for 5, 7 and 10 MHz could be used, with third harmonic resonances at 15, 21 and 30 MHz, giving six frequencies reasonably spread across the desired band.

Although this is a simple solution, it mostly results in slow frequency changing and discontinuous coverage, which are unacceptable to many users, and more radical solutions have to be found. Modern adaptive array 'smart' antennas use numerous elements with switching and tuning by ATUs, both automatic, to obtain the desired frequency coverage. They rely on the advantages gained from lobe steering to overcome the deficiencies of the individual elements.

7.3 Log periodic arrays

To avoid switching and tuning altogether it is possible to connect a number of dipole antennas permanently to the feeder. This is the basis of the **log periodic array** antenna (Fig. 7.2) (Isbell, 1960). The log periodic array is quite a useful form of antenna and is not really very difficult to understand. The array consists of a series of dipoles of graduated length and spaced so that their tips fall on the outline of an isosceles triangle, laid on its side. All the dipoles are driven, but (most importantly) the feeder is transposed between them, so that each is driven in antiphase to its neighbours.

End-fire characteristics are obtained, with the main lobe of the polar diagram on the axis of the triangle, in the direction from base to vertex. In Fig. 7.1, five dipoles are shown, but any number is possible, although performance is poor with as few as three and the antenna becomes impossibly cumbersome with more than about 10. The length of each dipole is made a constant factor times the next shorter (a geometric progression), so if dipole (1) is of length L and

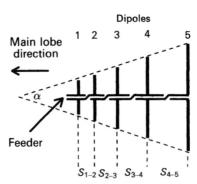

Fig. 7.2
A log periodic antenna.

distant d from the vertex, then the rth dipole has a length and position given by

$$L_r = L \cdot b^{r-1} \text{ and } d_r = d \cdot b^{r-1} \qquad (7.1)$$

where b is a constant.

To construct the antenna the values of s are required (as shown in Fig. 7.1). But

$$\begin{aligned} s_{(r-1) \to r} &= d_r - d_{r-1} \\ &= d \cdot b^{r-2} \cdot (b-1) = b^{r-2} \cdot (b-1) \cdot \frac{L \cdot \tan(\alpha/2)}{2} \end{aligned} \qquad (7.2)$$

Choosing vales for α, the number of elements, b and L, eqns (7.1) and (7.2) make it possible to complete the design of the antenna. How are these values chosen?

To understand this we must look briefly at how the antenna works. Since each dipole has a neighbour in antiphase, the fields that each would produce tend to cancel so that there is relatively little radiation. The exception is a dipole at or near resonance, when the current is increased approximately by the Q-factor, making its field far more than that of its neighbours and resulting in substantial radiation. Provided that the spacing is suitable, the non-resonating dipoles on either side will have a beam forming action, somewhat like the directors and reflectors in a parasitic Yagi array. Thus the antenna only works in the frequency range from that at which the longest dipole resonates (at the lowest extreme) to the resonant frequency of the shortest dipole (at the high end). The latter sets the value of L which must be a half wavelength at the highest frequency of interest.

The value of b sets how rapidly the length of successive dipoles increases. If it is too small the increase in length over a reasonable number of elements will be small and hence so will the frequency range covered, but if it is too large the beam forming, and hence the antenna gain, will be poor, because neighbouring elements will not have the right relative lengths for the best phase relationships. In practice, values ranging from 1.05 to 1.4 give favourable designs, the smaller figure giving best gain and the larger best bandwidth for a given number of elements; as usual there is a compromise between these factors. The vertex angle α sets the distance between elements (eqn (7.2)); if it is small it makes for a very long and unwieldly array but if too large it again compromises the gain and polar diagram. Values between 15 and 40° result in a usable range of designs.

The bandwidth of the array is set by the ratio R of maximum to minimum dipole lengths (corresponding, as already stated, to the ratio of minimum to maximum resonant frequencies). Considering eqn (7.1) this is

$$R = \frac{L \cdot b^{n-1}}{L} = b^{n-1} \tag{7.3}$$

Since the number of dipoles is usually set by considerations of cost, and is therefore fixed in advance, this equation is often used to find b given the required value of R (determined by use considerations). Of course, b must fall within the acceptable range, and if it does not then either n or R must change.

Log periodic arrays are occasionally seen as UHF television antennas, and they have had quite extensive use in the HF band. In that service, almost always horizontally polarized for good HF sky wave transmission, they are often sited on top of a mast (to get them away from the ground plane and environmental clutter) and commonly use a rotary mount so that they can be aimed on any desired compass bearing. Typical performance is for HF arrays to cover a frequency range of up to 1.5:1 and achieve power gains from 5 to 10 dBs (relative to isotropic) for from five to seven elements, respectively (Orr, 1987). Using Kraus's approximation, the lobe widths are of the order of a few tens of degrees. Although bulky, they are smaller and cheaper than a number of switched Yagi arrays (which may be the alternative) and have the advantage of giving continuous frequency coverage.

Log periodic arrays are often described as 'self-scaling'. This is simply to make the point that all the dimensions are proportional to L, so that by scaling this up or down the whole antenna will grow or shrink proportionately, scaling the operating frequencies also by exactly the same factor but leaving all other characteristics of the antenna, such as R and the polar diagram, completely unchanged.

Finally, why is it called a log periodic array? Simply because some of its parameters show a periodic dependence on the log of frequency.

7.4 Non-resonant antennas

In log periodic arrays the dipoles still resonate and in consequence antenna design gets difficult if the frequency range much exceeds 1.5:1. However, by avoiding resonance in an antenna altogether a much wider band can be covered. The resonance conditions for the dipole arise because of reflections at the conductor ends, giving rise to standing waves (Section 3.4). By scrupulously avoiding reflections, and hence standing waves, antennas can be built entirely without resonances. However, reflections occur not only at the ends of wires but at any sharp discontinuity (for example, if a wire suddenly changes direction or passes close to another conductor). The antennas must therefore be designed with great care, and they are inevitably long because of the need for only very gradual changes over their length. For this reason they are often called **long-wire antennas**.

HF (3–30 MHz) is the natural home of the long-wire antenna, and their use outside this band is negligible. In the MF band and at lower frequencies they would be quite impracticably large, whilst at VHF and above other antennas, such as switched Yagi arrays or log periodics, offer better performance whilst being cheaper and more compact. Although many long wire antennas will be found in use all around the world, their popularity is waning in the face of competition from 'smart' adaptive arrays, which give improved performance (including interference rejection by null steering) at no greater cost than the group of long-wire antennas needed to provide omnidirectional coverage.

A common long-wire antenna form is the **inverted-V** (Fig. 7.3). Variants on this basic type exist, including some which are fed at the

110 Radio Antennas and Propagation

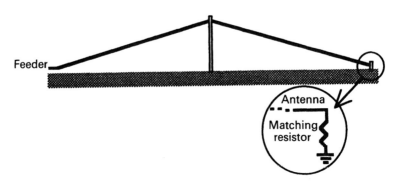

Fig. 7.3
An inverted-V long-wire antenna.

apex rather than at one end, as shown here, but they differ little in performance. As the name suggests, a single mast supports an inverted V-shaped wire antenna, which is long compared with its maximum height. Provided that the gradient of the wire is small enough (less than 15%) no significant reflection occurs at the point where the feeder joins the antenna, or with the change of wire direction at the mast top. As the wire and its image in the ground plane move apart the chance for radio quanta to escape increases (see Appendix), and dominates for separations which are large compared with the near–far transition radius $\lambda/2\pi$, that is an antenna height above ground exceeding 0.08λ. Given the limited gradient on the wire this implies a length of several wavelengths. At the end of the antenna, where it approaches the ground, a matching resistor is connected, chosen exactly to absorb power not radiated by this point, so there is no reflection here either. This resistor may absorb up to half the power flowing in from the feeder, so the efficiency of an antenna of this type is not good. However, with no reflections there are no standing waves, and hence no resonance effects. The antenna works entirely by means of the travelling waveform which enters at the feeder and is absorbed at the termination.

Each wire has a polar diagram approximately symmetrical around the wire. As might be expected, there is a null along the axis of the wire but with major lobes to either side, which approach nearer to

Wide-band antennas

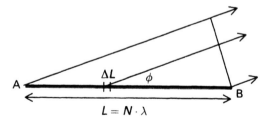

Fig. 7.4
The inverted-V antenna seen from above.

the wire the longer it is. The main lobes of the two wire sections align with the horizontal forward direction to give an end-fire overall characteristic. Since the electric field is between the wire and the ground, the antenna is vertically polarized.

To estimate the performance of an antenna of this type, consider it in plan view (Fig. 7.4). The radio frequency energy is guided along the wire of length $L = N \cdot \lambda$ from A to B, where it is absorbed without reflection. Since the wire is thin, the velocity of propagation along it is not significantly different from the free-space velocity c. However, each small element ΔL of the wire radiates some radio photons. We consider the signal received at a remote point P (not shown), which is so far distant that the radio quanta may be considered to take parallel paths to it. The direction of P is at an angle ϕ to the antenna.

In the case $\phi = 0$, with P on axis, considering the time radio quanta take to reach P from the point A, it will be seen that it does not matter at what point the energy ceases to be guided along the wire and is radiated, since the velocity and direction is the same in both cases, so all radio quanta arrive at P after the same time from A. For non-zero ϕ this is not so, because photons emitted from near point B have had a longer path than those from A. Consider all the small elements of the antenna and assume that all emit radio quanta equally. If each element ΔL produces a voltage in the receiving antenna at P equal to $\Delta E \cdot \cos(\omega t + \psi)$ and corresponds to a phase increment $\Delta \psi$, then if Ψ is the maximum phase shift for photons originating at end B, each small element produces an incremental signal amplitude

$$\Delta E = E_0 \cdot \frac{\Delta \psi}{\Psi}$$

where E_0 is the maximum amplitude.

$$E = \int_{\text{all } \Delta L} \cos(\omega t + \psi) \cdot dE = \frac{E_0}{2\pi} \int_0^{\Psi} \cos(\omega t + \psi) \cdot d\psi$$
$$= \frac{E_0}{2\pi} [\sin(\omega t + \psi)]_0^{\Psi}$$

If the maximum phase shift Ψ is 2π, that is if the path difference between signals received from the extreme ends of the antenna is one wavelength, the received voltage must be zero. Any trigonometric function is the same when its angle is advanced by 2π, so the upper and lower limits are identical. What this means in practice is that the signal at P will fall to zero when as many elements like ΔL are radiating in the positive half cycle as are doing so in the negative half cycle.

This results in an angle for the first zero Φ where

$$L(1 - \cos \Phi) = \lambda$$

so

$$\Phi = \cos^{-1}\left(1 - \frac{1}{N}\right) \tag{7.4}$$

Provided that the shape of the polar diagram is not too different from a cosine function, the angular distance from maximum to first zero is close to the distance between half-power points, so Φ approximates to the main lobe width.

A plot of the lobe width and the power gain (obtained from Kraus's approximation) is shown in Fig. 7.5. These results must be regarded as ball-park figures only, since the assumption that all elements of the antenna radiate equally is quite a crude approximation, also the

Wide-band antennas

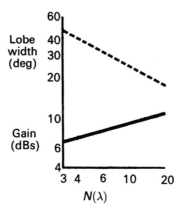

Fig. 7.5
Estimated main lobe width and gain for an inverted-V.

effects of wire gradient on the horizontal component of propagation speed have been ignored.

Although the antenna functions without resonance over a wide frequency range, the length in wavelengths is proportional to the operating frequency, and the main lobe width varies correspondingly. This can be a limiting factor on the usable frequencies.

The inverted-V antenna is vertically polarized. What can be done if horizontal polarization is required? Obviously, a V-shaped antenna can be placed 'on its side', with the wires horizontal, and these are invariably driven at the apex. However, it is a disadvantage that to support such an antenna requires three masts. For just one additional mast it is possible almost to double the gain and aperture of the antenna, using a **rhombic** structure (Fig. 7.6).

Four masts support the antenna, isolated by mast-top insulators, and at the end remote from the feeder a matching resistor absorbs all the power flowing to it, and thus prevents reflection and resonance. As with the inverted-V, the usual constructional precautions must be taken to avoid reflections at any point on the antenna or feeder. If each side of the antenna (from feeder to matching resistor) is N wavelengths, the gain will be about 3 dB better than

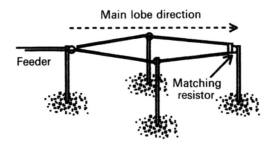

Fig. 7.6
A horizontally polarized rhombic antenna.

indicated in Fig. 7.5 and the angular width of the main lobe about half.

With four masts, the principal disadvantage of the rhombic antenna is cost, along with the large area it occupies, but set against this are its outstanding bandwidth and gain. For decades it was regarded as the 'Rolls Royce' of HF antennas, and major radio monitoring and marine radio-telephone sites often had as many as five or six, oriented to give coverage of all the continents, with (in Europe) two aimed across the Atlantic, for the North and South Americas. Recently, however, among the most sophisticated professional HF users the rhombic antenna 'farm' has been losing ground to active arrays.

Long-wire antennas are not the only non-resonant structures, although they have long been by far the most commercially significant. At frequencies too high for their use, particularly the very popular UHF band (300 MHz–3 GHz), an alternative has been proposed, based on equi-angular spirals (Rumsey, 1957). They have seen limited use in the UHF and SHF bands, but it is difficult to obtain a sufficient aperture with practicable designs, and are now facing competition from 'smart' adaptive array antennas in many applications.

Problems

1. A log periodic array is to transmit over the range 100–150 MHz. What are the lengths of the largest and smallest elements? [1.5 m, 1 m] If the antenna is to have the maximum practicable gain, what should be the number of elements? [9] What will be its length? [1.9 m] What value of gain would you expect? [>10 dB]

2. What are the principal advantages and disadvantages of non-resonant antennas? Describe two common types, one for vertical and one for horizontal polarization, indicating their technical characteristics and showing how the main lobe width may be estimated. What are the main applications of non-resonant antennas?

3. An MF receiving site operating from 1 to 3 MHz uses a single fixed antenna to receive vertically polarized signals arriving from directions ranging in angle over a total of 30°. What type of antenna is likely to be best for this service? [inverted-V] Illustrate your answer by a sketch. What should its dimensions be? [700 m long, mast 53 m high]

Chapter 8

Odds and ends

Some antennas defy easy classification, so in this chapter we will consider a number of types that fall outside the usual categories, some little more than curiosities but others having important applications in specialized niches. Specifically we will review magnetic, helical and near-field antennas.

8.1 Magnetic antennas

In Chapters 3 and 4 the theory of dipole antennas and arrays was developed from the properties of the short dipole or doublet. Properly speaking it would have been more correct (though hardly anybody ever does) to speak of an 'electric doublet' since a **magnetic doublet** is also possible (Fig. 8.1). A small loop carrying the radio frequency current has a pattern of electric and magnetic fields which exactly mimic those of the electric doublet, but with magnetic and electric fields interchanged. The two doublets are said to be **duals**, identical except for the interchange of electric and magnetic parameters. The equations that describe the magnetic doublet are identical with those of its electric counterpart except that voltage is replaced everywhere by current (and conversely), impedance by admittance and so on.

This 'Alice through the looking glass' world of the magnetic doublet can be worked out in just as much detail as its more familiar electric counterpart. The radiation emitted by the magnetic

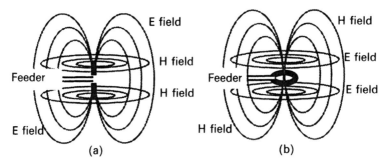

Fig. 8.1
An electric double: (a) in the form of a short dipole, and a magnetic doublet; (b) formed by a small loop.

antenna is polarized at right angles to the loop axis, in contrast to the electric dipole where the polarization is parallel to the antenna axis. To receive vertically polarized waves, for example, the axis of the loop must be horizontal. The polar diagram is the familiar 'doughnut' around the axis, so for a horizontal axis antenna it looks like a figure-of-eight in the horizontal plane. As for the equivalent circuit, this is now a parallel rather than a series resonant circuit (Fig. 8.2).

The admittance presented at the antenna terminals is

$$Y = j\omega C + \frac{1}{j\omega L} + G \tag{8.1}$$

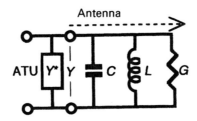

Fig. 8.2
Equivalent circuit of a magnetic loop antenna near resonance, with its antenna tuning unit (ATU).

In this case the antenna tuning unit (ATU) is an admittance in parallel with the antenna terminals

$$Y^* = -\left(j\omega C + \frac{1}{j\omega L}\right) = j\omega C\left(\frac{1}{\omega^2 LC} - 1\right) \qquad (8.2)$$

Provided that $1/LC$ is large compared with the square of the operating frequency this amounts to a capacitor of value given by eqn (8.2).

The exotic-seeming theory of magnetic antennas has exerted a fascination in the past, but the commercial use of loop antennas has been very restricted. In theory it is possible to join many such doublets together and have large resonant loop antennas, also loop antenna arrays (both driven and parasitic) can be built, and so on. In reality such things are exceedingly rare. Nevertheless, there is one application to which the magnetic antenna is ideally suited, and this is a universal one. Paradoxical as it may seem, antennas of this type are among the most commonly encountered. Virtually all portable MF (medium wave) and LF (long wave) broadcast receivers use a **ferrite rod antenna**, which is a form of magnetic loop antenna (Fig. 8.3). (The LF winding is omitted in the USA, where LF is not a broadcast band.) Ferrite is a type of ceramic which, although an insulator, has ferromagnetic properties, with a relative permeability of at least 100 and for some materials much higher.

This magnetic dual-band receiving antenna uses many turns in a winding for each band. The effect of multiple turns is the

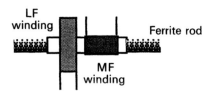

Fig. 8.3
A ferrite rod antenna as used in virtually all modern LF/MF broadcast receivers.

counterpart of a folded dipole in the electric case (where the dipole is effectively duplicated) but by reason of favourable geometry, the ferrite rod antenna is able to carry the idea much further. It raises the matching resistance greatly, and in this case also increases the inductance of the winding and hence L in the equivalent circuit of Fig. 8.2. This is further increased by introducing a ferrite core into the winding. In practice the ferrite rod is most commonly of circular cross-section and may be 1–2 cm in diameter. Its length varies from a few centimetres up to 30 or more, depending on the size that can be accommodated in the receiver cabinet. To reduce the self-capacitance, the winding is usually single layer at MF. For the LF coil this is not practicable because of the larger inductance required to achieve resonance at a lower frequency, and a multi-layer coil must be used, although it is often wound in 'basket' form to keep the capacitance down.

Considering eqn (8.2), if the inductance is large the ATU becomes simply a capacitor; in a broadcast receiver this is the tuning capacitor at the input port of the receiver. By making it variable (for example, using a voltage-controlled varactor) the condition of eqn (8.2) can be met over a range of frequencies covering the band, permitting the selection of a wanted radio station.

It will be recalled from the electric case (Section 2.3) that within the area in which induction (near-field) effects are dominant (the near–far transition radius) a radio quantum is very likely to be captured; introducing a ferrite rod greatly extends this area, depending primarily on the permeability and length of the rod. By locally distorting the waves associated with passing quanta, the ferrite rod thus greatly increases the capture area of the antenna, resulting in quanta being 'caught' which would pass by if the core were not present.

Ferrite rod antennas are cheap, effective and easily tuned. Always used for vertically polarized signals and therefore with the rod horizontal, they have a 'figure-of-eight' polar diagram as the receiver is rotated about a vertical axis, and an informed user can sometimes exploit the antenna nulls as a means of rejecting sources of interference. Almost universal in MF and LF broadcast receivers, so far they have not proved satisfactory in the HF and

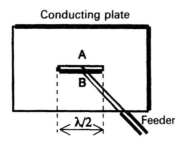

Fig. 8.4
A slot antenna for vertical polarization.

VHF bands because of lack of ferrite materials combining low losses and high permeability in the higher frequency ranges.

Ferrite antennas are almost never used for transmission because the ferrite material saturates at relatively modest levels of magnetization, which puts a very low limit on the current which can flow in the windings without causing a drastic fall in core permeability. For reception this is entirely unimportant, but in transmission it would limit the power levels the antenna could handle to the milliwatt range, depending on the antenna dimensions, which is useless for any but the shortest range applications.

A different type of magnetic antenna, occasionally seen, is the **slot antenna** (Fig. 8.4). A horizontal slot half a wavelength long is cut in a conducting plane, its width typically one-fiftieth of a wavelength. It is fed at two points A and B at the centre of the upper and lower edges of the slot. Current flows around the edges of the slot on either side of the feed points and with an opposite sense of rotation; thus when, for example, the magnetic field vector points forward from the left half it points backward from the right, so that the field takes a circular horizontal path. Since the magnetic field vector is horizontal, the electric component is at right angles to it, in line with the axis A–B. A horizontal slot thus produces a vertically polarized wave, and conversely.

As compared with an electric dipole, which has very similar electrical characteristics, this curious antenna has marked advantages only in the SHF and EHF bands, where dipoles become unacceptably fragile but the slot antenna can still be robust. It is sometimes chosen as a radiating element in aircraft, using the vehicle skin as the conducting plane, and there has been military interest in its use for antennas resistant to bomb-blast. Sometimes it is backed with a second conducting plane to form a plate antenna, giving it directionality, and arrays are possible, although never seen below SHF. At the shorter wavelengths it merits consideration particularly in constructing the active cells for use in adaptive arrays (see Sections 4.5 and 4.6) because it makes the front surface of the array a continuous conducting plane, to be fabricated in metal with consequent structural and protection advantages.

Slot antennas are also extensively used as vertical collinear arrays for high-power broadcasting transmitters (for example, television). A vertical slot, easily accommodated on a slim mast, radiates a horizontally polarized signal. Using horizontal dipoles would result in an unsatisfactory mechanical design, less able to survive the wind loading on a high mast.

8.2 Helical antennas

The advantages of using circular polarization of radio transmissions in some applications have already been noted. One antenna, the **end-fire helix**, naturally launches circularly polarized emissions (Fig. 8.5). If we imagine photons being launched and travelling to the right, those emitted at points such as A_1 and A_2 will be in phase if the time taken for a guided wave to travel around the helix between these points is the same as that for radio quantum to move through the space between them plus an integral number of wavelengths.

Considering one complete turn of the helix, this can be seen as equivalent to an inclined conductor wrapped around an invisible cylinder. The conductor is therefore the hypotenuse of a right-angled triangle of base equal to πD and height p. Hence we have the condition

$$\sqrt{D^2 + p^2} = p + n\lambda \tag{8.3}$$

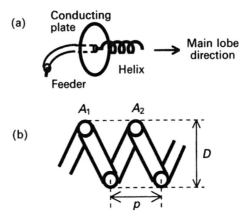

Fig. 8.5
(a) An end-fire helical antenna. The conducting plate may be a wavelength in diameter, and is distant $p/2$ behind the helix. (b) Helix dimensions.

where n is an integer, almost always 1, and λ is the designed operating wavelength.

Putting $n = 1$, this comes down to

$$\frac{D}{\lambda} = \sqrt{2\frac{p}{\lambda} + 1} \tag{8.4}$$

This condition must be met if all corresponding points on successive turns of the helix (like A_1 and A_2) are to radiate in phase. The relationship (which ensures maximum main lobe magnitude and hence maximum gain) is plotted in Fig. 8.6.

The bandwidth of a helical antenna is determined by the same considerations. When the condition of eqn (8.3) is not met corresponding points on successive turns of the helix no longer give rise to in-phase radiation. For a helix designed to have points on successive turns in phase for a wavelength λ (frequency ω), the two sources of radiation A_1 and A_2 (Fig. 8.5) will have a phase

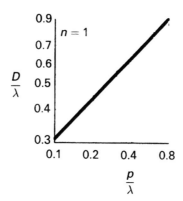

Fig. 8.6
Conditions on helix dimensions.

difference θ at wavelength λ' (frequency ω') where

$$\theta = \pm(\lambda' - \lambda)\frac{2\pi}{\lambda} = \pm\frac{2\pi(\omega - \omega')}{\omega'}$$

Assuming that the frequencies are not too different, the bandwidth of the antenna between points at which this angle reaches θ is given by the difference between these upper and lower limits, and is

$$\Delta\omega = \omega \cdot \frac{\theta}{\pi} \tag{8.5}$$

If we now calculate the value of θ which corresponds to a 3 dB reduction in the transmitted signal, we can obtain the effective bandwidth of the antenna. Taking points such as A_1, A_2 and so on, on each turn of the helix, each will contribute a component to the resultant transmission. Supposing that there are N turns, there will be N such components, spaced from each other by $\theta°$ (Fig. 8.7). The resultant is as shown and is the sum of the equal magnitudes of the individual vectors each multiplied by the cosine of the angle between the vector and the resultant. (Strictly, of course, one should write 'phasor' rather than 'vector'.) Thus, if each individual component is of magnitude e, from symmetry we may sum for

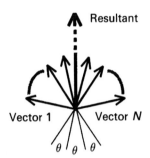

Fig. 8.7
Vectors resulting from successive turns at a frequency off the design value.

the right half of them only and double it, so the resultant is

$$e_r = 2e \sum_{r=1}^{r=\frac{N}{2}} \cos(r - \tfrac{1}{2})\theta \tag{8.6}$$

The summation can be evaluated as

$$S = \sum_{r=1}^{r=\frac{N}{2}} \cos(r - \tfrac{1}{2}) \cdot \theta = \Re\{\sum \exp[j(r - \tfrac{1}{2})\theta]\}$$

which can be summed as a geometric series.

Hence

$$S = \frac{\sin\left(N \cdot \frac{\theta}{2}\right)}{2 \sin \frac{\theta}{2}}$$

Note that if $\theta \to 0$, $S \to \frac{N}{2}$ and further $e_r = 2e \cdot \frac{N}{2} = Ne$ as expected when the antenna is working at its design frequency. At what frequency is the resultant reduced by 3 dB, the power halved and so the voltage reduced by $\sqrt{2}$? Evidently this happens when

$$\frac{N}{\sqrt{2}} = \frac{\sin\left(N \cdot \frac{\theta}{2}\right)}{\sin \frac{\theta}{2}}$$

The RHS numerator is near to unity for likely values of the variables N and θ, so to good approximation

$$\theta = 2 \sin^{-1}\left(\frac{\sqrt{2}}{N}\right)$$

Combining this result with eqn (8.5) above

$$\frac{\Delta\omega}{\omega} \approx \frac{2}{\pi} \cdot \sin^{-1}\left(\frac{\sqrt{2}}{N}\right) \tag{8.7}$$

This relationship is shown in Fig. 8.8. As the number of turns increases the bandwidth falls, but this is still an antenna of moderately wide bandwidth. Note that this curve assumes that operation of the antenna to the -3 dB point is acceptable, but (to take the transmitting case) half the transmitter power is lost at this point. If this loss is unacceptably large a narrower bandwidth must

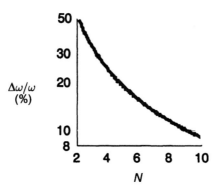

Fig. 8.8
Variation of helix bandwidth with number of turns.

be accepted, by replacing the $\sqrt{2}$ term in eqn (8.7) by a number closer to unity.

Other parameters of the end-fire helical antenna array can be calculated similarly. For a single turn we would expect the aperture to be close to $(\pi D^2/4)$, but to obtain the aperture of the helix we cannot simply multiply by N because it also depends on p. If the pitch is small the individual turns are too close to be regarded as independent in capturing passing photons. More detailed analysis gives the aperture as

$$A_h = \left(\frac{\pi D^2}{4}\right) \cdot N\left(\frac{15p}{\lambda}\right) \tag{8.8}$$

The power gain follows, dividing by $\lambda^2/4\pi$ (the aperture of an isotropic radiator) and is

$$G_h = 15\pi^2 \frac{NpD^2}{\lambda^3} \tag{8.9}$$

A good 'rule of thumb' for the gain (invariably marginally lower than eqn (8.9) indicates) is that it is 140 times the volume of the helix *expressed in wavelengths.*

Similarly the matching impedance is

$$Z_h = 140\frac{\pi D}{\lambda} \tag{8.10}$$

This will be recognized as 140 times the circumference of the helix *expressed in wavelengths.*

Orr (1987) quotes a typical design for an end-fire helix for the UHF band. End-fire helical antennas are useful at VHF and UHF by virtue of their good bandwidth and aperture. They are not used at lower frequencies because of their physical size, whilst at SHF and above, compared with other practicable antenna types, they do not give a large enough gain to be attractive.

If a helix has D small compared with a wavelength the axial radiation disappears and the antenna radiates broadside. Such a helix is often mounted vertically over a horizontal conducting plane, when it acts much like a conventional whip antenna, except that it resonates roughly when the length of conductor reaches a quarter wavelength, at which point the length of the helix is considerably less. This can be a cheap way to avoid using an ATU, although in aperture nothing is gained over a straight whip. Polarization from an antenna of this kind is linear and parallel to the axis of the helix.

8.3 Near-field antennas

Although most antennas are used for transmission over considerable distances, there are other applications where the radio signal is required to be effective only over a very short range. In this case, to conserve the electromagnetic spectrum by minimizing the area of possible pollution and therefore minimize the chance of interference to or from distant users, a near-field antenna may be chosen. Consider two magnetic doublets driven in antiphase (Fig. 8.9). It is obvious that at a distance much larger than d (itself assumed small compared with a wavelength) the fields of the two will be very nearly equal and opposite and will therefore cancel. If the distance between the doublets is similar to their diameter the antenna approximates to a **quadrupole**, which has near field but almost no far field. Many quanta are emitted by the antenna but they have a high probability of being recaptured within the near-field region, and few escape.

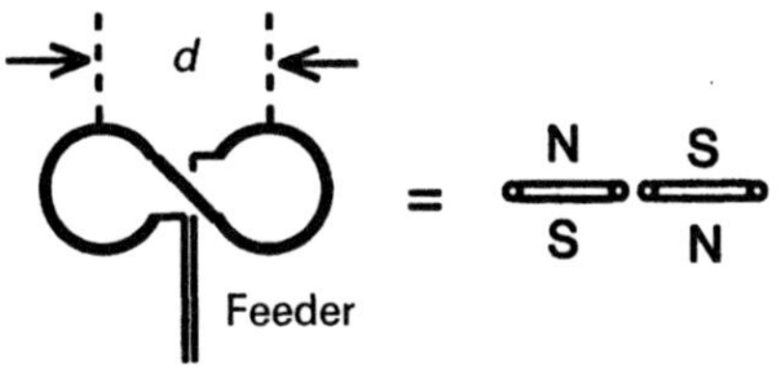

Fig. 8.9
A quadrupole, approximated by two antiphase magnetic doublets.

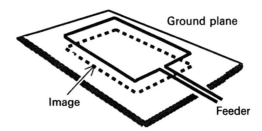

Fig. 8.10
A horizontal loop over a conducting ground plane approximates to a quadrupole. It has almost no far field.

A practical approximation to this arrangement is often built using a horizontal magnetic loop antenna over a ground plane (Fig. 8.10). The loop is a magnetic doublet but its antiphase image, formed below in the conducting surface, gives the antenna properties which approximate to a quadrupole. This type of antenna would be used when very localized communication is required, virtually limited to the area within the loop and a short distance outside it, comparable with the distance between the loop and its image. One very common application is as the vehicle detector for automatic traffic lights, where the absorption of energy by a vehicle when it is over the antenna is the indication. The loop is buried just below the road surface; the highway paving is insulating, so the image is formed in the earth some tens of centimetres lower down. Sometimes similar antennas are used to signal to road vehicles, or to interrogate them in road pricing schemes. Another application for antennas of this type is in very localized broadcasting schemes, such as on hospital or university campuses, although in this case the antenna loop is above the surface to extend the range somewhat. Note that in order to act as a magnetic doublet the loop must be small in size compared with a wavelength, so antennas of this type are chosen only at MF and lower frequencies, where the wavelength is large enough to give a useful coverage area. Traffic signal loops operate at LF.

For higher frequencies an alternative approach is needed. This is

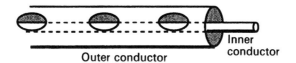

Fig. 8.11
A leaky feeder near-field antenna.

provided by the use of **leaky feeders**. In the example shown (Fig. 8.11) a coaxial cable has a series of holes punched in its outer conductor. Other versions of the same idea use lengthways slots in place of round holes, or even just a very loosely woven metal braid as the outer conductor.

In the transmission mode, radio quanta leak out through the holes, but very few escape to form a far field due to destructive interference from adjacent holes. In free space, very approximately, the signal is found to fall off as a negative exponential function of distance from the cable, measured in wavelengths. Thus the received power

$$p_R \approx k_1 p_T \exp\left(\frac{-k_2 r}{\lambda}\right) \quad (8.11)$$

where k_1, k_2 are constants.

The range to which such an antenna functions depends on the transmitted power and on the receiving antenna location and receiver sensitivity, but can extend to a few wavelengths from the cable. However, free-space characteristics are rarely important. The application of this kind of antenna is most particularly in mines and tunnels, where far-field propagation is unsatisfactory due to reflection and absorption at the tunnel walls (Delogne, 1982). It is also sometimes used inside buildings, where the ability to confine the coverage to certain areas can be an important advantage. They are used principally in the VHF band (30–300 MHz) where they give excellent coverage quite a few metres either side of the cable. They can also be used at UHF (300 MHz–3 GHz) but as the wavelength grows shorter the working range falls, until it gets too small to be useful.

In the usual enclosed environment, such as a tunnel, eqn (8.11) does not apply because quanta are scattered back from the walls. The picture is complicated. Quanta escape through the holes, but subsequently behave in a variety of ways. They may travel along the tunnel by a series of reflections from the walls, but they may also be guided along the surface of the cable. In that case some are lost or absorbed but their numbers are sustained by leakage from subsequent holes, a mode known as Goubau propagation. If the cable is near the tunnel walls (the usual case) quanta will travel along the cable in this way but concentrated mostly in the space between the cable and the wall, in consequence of the strength of the field there. Bends, supporting hardware or solid objects very near to the cable will cause the quanta to be reflected out into the space of the tunnel. Signal amplitudes are Rayleigh distributed (see Section 12.5) about a mean value which is fairly constant across the tunnel. The ratio of the mean received power to that in the cable is known as the coupling loss, and is dependent primarily on the distance between the holes, their shape and size and the aperture of the receiving antenna. In practice typical values are found in the shaded area of Fig. 8.12.

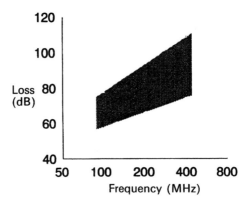

Fig. 8.12
Leaky feeder coupling loss.

Due to leaking radio quanta, the losses of a leaky feeder are much higher than those of a normal coaxial cable, the radio power level falling markedly along the cable away from the driven end. Sometimes booster amplifiers are used at intervals along the cable to overcome this problem.

Leaky feeders are just as useful for reception as for transmission, their range being much the same in both cases. Many installations use the same leaky feeder for both functions at different frequencies (or less commonly in different time-slots), enabling a base station connected to the leaky feeder to communicate with mobile two-way radios within its range.

Although a coaxial leaky feeder is shown in Fig. 8.11, parallel line feeders are also sometimes used. These always leak a little, the extent increasing the larger the spacing between the wires (see Appendix). By using small widely spaced conductors a useful near field is established. The disadvantage of the parallel wire feeder is that both conductors are at a potential to ground ('live') so neither can be attached to a structure, such as a building, without disturbing the electrical characteristics of the feeder. They are also much more affected by dirt or water. One or two railway tunnel installations have used the track as a parallel line leaky feeder (but at HF).

Problems

1. An AM broadcast receiver covers the range 600 kHz to 1.5 MHz and uses a ferrite rod antenna. The input port of the receiver may be represented as a fixed capacitor of 30 pF in parallel with a varactor (variable capacitance diode) having a maximum-to-minimum capacitance ratio of 10:1. What must be the upper value of the varactor capacitance if it is to cover this range? [420 pF] With the ferrite rod chosen for the antenna, a winding of 10 turns produces an inductance of 12.8 µH. How many turns will be required for the antenna? [35] (Hint: inductance is proportional to the square of the number of turns.)

2. Compare crossed Yagi arrays with an end-fire helix for launching circularly polarized transmissions. Why is the latter preferred? An end-fire helix at 300 MHz has a diameter of 0.5 m, a pitch of 0.25 m and eight turns. What will be its matching resistance and gain? [220 Ω, 55× or +17.4 dB] Estimate the main lobe width. [from Kraus's approximation, 30°]

3. A leaky cable radio transmitting system has a straight run in a non-reflecting environment. The system is engineered for 150 MHz and on installation test gives a received signal which is just acceptable 2 m from the cable with a minimum transmit power of 0 dBm and at 4 m with a power of +10 dBm. What transmitter power (in watts) is needed to give coverage to 8 m with a 10 dB safety margin? [10 W]

CHAPTER 9

MICROWAVE ANTENNAS

The term **microwave** is not precise. Many take it to comprise the SHF band (3–30 GHz) and EHF band (30–300 GHz) both together, whilst others would prefer to describe the EHF band as **millimetre wave**. A few use the term microwave to refer also to the upper part of the UHF band (300 MHz–3 GHz), say above 2 GHz. Although microwave products amount to less than 3% of the total radio market they have received much attention, partly because they often lend themselves to very elegant mathematical analysis, but also because some practical problems can be solved in no other way. Microwaves define a small niche market, but it is an important and growing niche.

So far as antennas are concerned, practice begins to change above about 1 GHz and by 3 GHz things are radically different, so although vague in definition, 'microwave' does represent a significant distinction. No similar discontinuity of design practice occurs at still higher frequencies, even though there is a progressive change in preferred antenna types. This chapter will therefore concern itself with antennas for 3 GHz and above, whilst recognizing that some of the factors to be described have their effects at somewhat lower frequencies also.

9.1 Microwave technology

The technology and form of microwave antennas is inextricably tied up with the way in which microwave radio energy is generated in the transmitter, and also how it is received. This underwent a profound change in the last half of the twentieth century.

The substantial use of microwave radio began in World War II (1939–45), primarily for radar, the successful development of which undoubtedly had a major influence on the outcome of that war. At that time the only active devices available for use in electronics were thermionic valves (or tubes), solid state electronics not yet having appeared. Special valves were developed for SHF use, notably the magnetron as a transmitting device and small klystrons for use in receivers. Later other thermionic valves, such as gyrotrons and ripple tubes, were developed to provide high power at EHF.

These devices, like all thermionic valves, were costly, large, fragile and very power hungry. It was therefore impossible to build the transmitters or receivers in a form suitable for mounting in close proximity to the antennas. Equipment had to be housed remote from the antennas, so long runs of feeder could not be avoided. However, the coaxial or twin-wire cables, universally used as feeders at lower frequencies, have high losses in the microwave bands, so it proved necessary to use **waveguide** runs in their place. Waveguides are long metallic tubes, usually rectangular in cross-section, in which (to take the case of transmission) radio quanta are launched at one end, at a point very close to the transmitter. Reflecting with little loss from the conducting walls of the tube, the photons are guided along the tube to the point at which they are to be emitted into space, which may be tens or even hundreds of metres away. The design of systems using waveguides was the object of much study during the middle of the twentieth century, and excellent books on the topic exist (Baden Fuller, 1990).

This technology was profoundly influenced by the fact that the components employed, such as valves, waveguides and other transmission lines, were of the order of size of the wavelength of the radio energy concerned, or larger. Thus it was impossible to assume that circuits were composed of well-defined components, such as resistors or capacitors, connected by wires which themselves had negligible impedance or admittance to ground, and both travelling and standing waves were likely to be present. Circuits of this kind are called **distributed circuits,** to distinguish them from **lumped-constant circuits,** familiar at lower frequencies, where all components are small compared with a wavelength. In the 1940s and 1950s an elegant theory of distributed circuits was worked out, many suitable waveguide components were developed, and it proved possible to perform all the functions required in receivers and transmitters this way.

Although they were successful and made possible the early exploitation of microwaves for many purposes, such waveguide-based systems had serious drawbacks. Because the inner surfaces of waveguides had to be flat to within a small fraction of a wavelength they were expensive to manufacture, and had to be relatively thick walled to ensure that the tolerances were maintained in use. Joints between components had to be very carefully made to avoid reflections and consequent standing wave effects, so the components had to have rigid flanges which could be bolted or otherwise fastened together with great precision. All this made microwave equipment of that time heavy, bulky and, above all, very expensive. The effect was to limit the exploitation of microwaves to premium applications, such as radar and high-capacity point-to-point links, where the cost could be justified.

By the end of the twentieth century all this had changed. Microwave solid state active devices became available at increasingly economic cost and thermionic valves (tubes) came to be used only in specialized applications. Before long magnetrons were rarely to be found anywhere because of their very poor radio spectrum utilization (except in the kitchen, as power sources for microwave ovens), whilst klystrons, though in very large sizes, appeared only in high-power transmitters (for example, as used in television broadcasting). The semiconductor devices, principally versions of bipolar and field-effect transistors, displaced all others in virtually every application. At first able to handle only lower powers, they rapidly improved, with ever higher powered devices becoming commercially available at economic cost. Because the active regions of solid state devices are very small compared with a wavelength, this made possible two important developments: first microwave integrated circuits (**MICs**) and later microwave monolithic integrated circuits (**MMICs**, pronounced 'mimics').

MICs are formed on an insulating substrate, most commonly a thin ($\approx$1–3 mm) sheet of high alumina ceramic, chosen both for its excellent low-loss electrical characteristics at high frequencies and for its good thermal conductivity, which facilitates heat dissipation from components. The substrates are typically rectangular and a few centimetres on a side. The circuit is formed by **thin film** technology. Thin metal or insulating films are deposited on the surface using vacuum evaporation or sputtering, and the deposited film is then shaped to the desired pattern by masking and selective etching. Often several layers are deposited and shaped to create the required circuit. Other components, which cannot be formed in this way, such as transistors and larger capacitors, are then attached to the circuit at the appropriate points, usually by soldering. Inductors can also be attached, but in view of the small values required at these frequencies they are mostly formed directly as small flat spiral conductors on the insulating substrate. Although the component sizes in MICs can be kept small enough for conventional lumped-constant circuit design techniques to be used, it is also possible, on slightly larger substrates, to incorporate lengths of transmission line, and to design using distributed constant approaches.

MMICs, by contrast, are actually fabricated entirely on a semiconductor chip, either silicon or gallium arsenide (which has advantages at higher frequencies due to the high mobility of electrons in this material). Components are formed by diffusion or ion implantation, in exactly similar ways to those adopted for circuits at lower frequencies. MMICs are much smaller than MICs, typically about a centimetre square or less, and generally do not use distributed circuit techniques, except at the highest EHF frequencies. Sometimes a hybrid type of circuit is seen, with several MMICs (along with other components) mounted on a MIC.

This dramatic change in technology, between waveguide circuits and MICs or MMICs, had a number of important results. The new microwave circuits use less power and are smaller, lighter, more rugged and above all far cheaper than the old. Whereas waveguide circuits were assembled by skilled technicians, MICs and especially MMICs can be mass produced, largely on automatic machinery. For these reasons waveguide circuits are no longer used at all in professionally designed receivers (even for production in small volumes), and not in transmitting equipment either, except in the highest power applications where their potential to handle megawatts is valued. By contrast, the low cost of the new technology and its adaptability to volume production has opened up large new markets to microwave solutions where they would previously have been uneconomic (for example, satellite television reception). Small, compact equipment can often be mounted in, or very near, the antenna structure, avoiding the need for costly long microwave feeder runs. All these points need to be kept in mind when considering the design of modern microwave antennas.

9.2 The basic problems of microwave antennas

Why should microwave antennas be different from those at much lower frequencies? Everything that is special about them follows from the very short wavelengths in these bands, ranging from 10 cm at 3 GHz, through 1 cm (10 mm) at 30 GHz down to 1 mm at 300 GHz. Resonant antenna lengths therefore become very small. A half-wave dipole is no longer than a person's thumb at 3 GHz and would be only just noticeable to the eye at 300 GHz (beyond which the atmosphere absorbs strongly, making the use of radio difficult anyway).

Bearing in mind that the aperture (capture area) of a simple resonant antenna is some constant multiplied by the square of its length, it will be appreciated that the aperture rapidly declines below usable values for simple antennas of this kind (Fig. 9.1). Practical applications for antennas with apertures less than, say, a tenth of a square metre are very few. Going to three-half-wave structures (rather than half-wave) can be a help, but only extends the usefulness of simple antennas moderately and introduces other problems, particularly restricted bandwidth due to higher Q-factor.

What is necessary in the microwave region is to have antennas of large aperture, which will therefore also have a large power gain (since this is the ratio of the aperture to that of an isotropic antenna). This unavoidably means that they will have a very narrow main lobe, which may be an advantage or a disadvantage, depend-

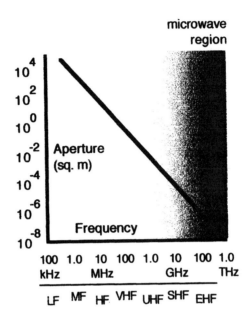

ing on the application. However, meeting this requirement is no easy matter, not least because the fabrication of microwave antennas itself presents some difficult problems.

Because of the very small wavelengths in these bands, it is necessary for antenna structures to be made far more precisely than at lower frequencies. For example, a parabolic reflector might deviate from its correct shape by 3 mm and this would only amount to an unimportant 1% of a wavelength at 1 GHz, but at 25 GHz it would be a quarter of a wavelength, causing serious phase changes in the wave functions of the radio quanta reflected at this point, relative to their design value. Similarly, a hole in a metal plate which is very small compared with a wavelength has a negligible probability that a radio quantum will pass through it, but as the frequency rises and the wavelength drops the probability that photons will 'leak' through holes steadily increases. So, in practice, whereas at 1 GHz a reflecting surface may be made of mesh and fairly crudely formed, at 25 GHz it will be unsatisfactory unless it is solid, and machined or cast to tight limits. Antenna construction for the microwave bands is mechanically more exacting than at lower frequencies, and hence more expensive.

A further difficulty is that dipole and other wire or rod antennas must have conductor widths which are small compared with their length, to give one-dimensional current flow, which means that in the microwave bands they become very narrow indeed, and are unacceptably fragile unless supported in some way. Thus at 300 GHz, a dipole would have to be constructed of wire only a few tens of microns thick, which is entirely impracticable if free-standing.

Even assuming that all these problems can be overcome, the result is a typical microwave antenna which may have a large gain and aperture but for which (recalling Kraus's approximation) the width of the main lobe must be correspondingly narrow. In these bands omnidirectional antennas are out of the question, and we are exclusively concerned with 'pencil beams' of radio quanta aimed at well-defined targets or receiving antennas. In consequence, for applications where this is just the characteristic that is most desirable, these bands are preferred. In the terrestrial radio service they are used for point-to-point links and radar systems, for both of which a narrow main lobe is essential. The omnidirectional antennas needed for terrestrial broadcasting would have hopelessly low gain, so these bands are at present used for broadcasting only from satellites, which are so far distant from us (about 36 000 km for synchronous orbits) that a narrow lobe still covers a considerable part of the Earth's surface. An antenna with a 1° main lobe width which is in synchronous orbit covers a circle on Earth over 500 km in diameter at the equator and a larger oval 'footprint' at higher latitudes.

9.3 Microwave array antennas

The simplest microwave antennas are constructed from arrays of driven dipoles (Fig. 9.2). To obtain a reasonable aperture a large number of dipoles are required; 50 are shown here but the number

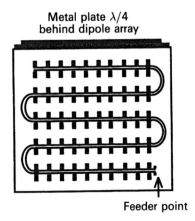

Fig. 9.2
A driven dipole array using etched-foil dipoles on an insulating substrate.

can run into many hundreds. Because the dipoles are necessarily very narrow to ensure approximately one-dimensional current flow, they are supported by an insulating substrate. The design of the array is exactly the same as described in Chapter 4. The dipoles are all mounted on a common feeder, care being taken with the connections between dipoles, since they must all radiate in phase, yet the spacing between them is considerably less than a wavelength. If the feeder is transposed between successive dipoles the necessary distance is reduced to $\lambda/2$, and this can be accommodated by a meander to ensure correct phasing. Care is also needed with the lengths of connections at the ends of rows.

At the lower microwave frequencies the etched-foil technique is used. A thin insulating board, usually of plastic-bonded glass fibre material having relatively low dielectric losses at microwave frequencies, has copper foil bonded to its surfaces. This is selectively etched on both sides to give the required conductor pattern using photolithography in exactly the same way as for printed circuit boards, and finally through-connections are made where necessary. Often the finished pattern is electroplated to improve conductivity and durability, using either silver or gold. This technique of manufacture lends itself to low-cost, large-volume production. At higher frequencies losses in the plastic substrate become significant and above 20 GHz the antenna may be fabricated by thin film technology on an alumina substrate, in the same way as an MIC; however, this is more expensive and does not lend itself well to antennas of large aperture.

To give the polar diagram a single main lobe, a metal plate is normally placed behind the plane of the dipoles, separated from them by a quarter of a wavelength (making due allowance for the refractive index of any insulating material that is present). The theory of the antenna is just the same as that for the plate antenna described earlier in Section 6.1. Flatness of the dipole plane is critical, as is its spacing from the reflector plate. Any flexing by more than a small fraction of a wavelength will have a major effect on the polar diagram, so the structure must be made adequately stiff, and this limits practicable sizes. Due to the directionality produced by the reflector plate, antennas of this type have an aperture just a little larger than the area enclosing the dipoles (provided that the number of dipoles is large) and if the area of the array is A the gain relative to isotropic is close to

$$G = \frac{4\pi A}{\lambda^2} \qquad (9.1)$$

For example, an array 30 cm square operating at 10 GHz has a gain of just over 30 dB.

These structures offer an economical way of building a microwave antenna at frequencies up to about 30 GHz, beyond which it is increasingly difficult to construct an antenna of consistent performance with a useful aperture.

However, if an array of this kind is to be built, it makes a great deal of sense, at least in the more sophisticated applications, to convert it into an active adaptive array, as described in Sections 4.5 and 4.6. Increasingly, this is the trend with advanced antennas for the higher frequencies. It brings a number of advantages:

- Elimination of all microwave feeder losses, since the radio frequency circuits are integrated in the antenna.
- Lobe steering, which is particularly valuable when communicating with mobile users since omnidirectional antennas are impracticable.
- Null steering when receiving, which can greatly improve the wanted signal-to-interference ratio.

Going beyond these 'printed circuit' arrays, however, it is possible to construct dipole arrays on the surface of a semiconductor slice. Bearing in mind that silicon slices are fabricated with a diameter as large as 30 cm (leading to antenna aperture around 0.07 sq m, allowing for a reflecting back plane) and given that using integrated circuit fabrication technology it is easily possible to define features to less than 1 µm, this has obvious advantages for the higher frequencies. Even at 300 GHz, a resonant dipole, say 0.75 mm (750 µm) long, can have an acceptable length–width ratio and be well-enough defined to give highly predictable performance. Its aperture alone would be very small, but it is perfectly possible to fabricate and interconnect many hundreds or thousands of dipoles on a single slice, and even to interconnect many slices. The first arrays of this kind were fabricated on insulating GaAs substrates, but these are expensive and available only in smaller sizes. Later fabrication on high-resistivity silicon was reported. Once again, fabrications of this kind are backed with a conducting plate or film to ensure that there is only a single main lobe, perpendicular to the surface of the slice (Fig. 9.3). The aperture of such an array is approximately the area of the slice (when used with a back reflector, otherwise half that) and if the diameter of the slice is D the gain relative to isotropic is nearly

$$G = \left(\frac{\pi D}{\lambda}\right)^2 \tag{9.2}$$

However, if an antenna array is to be formed on a semiconductor slice there is no disincentive to making it an active adaptive array. Although it is possible to design an active antenna array on a slice for both transmission and reception, switching the electronics connected to each elementary dipole when the mode is to be

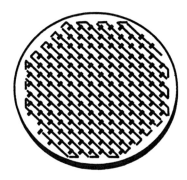

Fig. 9.3
An array of 160 dipoles formed on the surface of a semiconductor slice.

changed can present complications. If full duplex operation is required two slices may be used, one carrying the receiving antenna and the other the transmitter.

Because in the transmitting mode each dipole has a separate radio power source of its own, very many transistors or similar devices contribute to the total output power, each one has a relatively very small power rating, which eases device design. The effective area of the slice driving each dipole is small, so lumped-circuit design can be used. For the same reason, heat generated as a result of losses in the electronics driving each dipole is distributed uniformly all over the slice, which makes cooling easier.

9.4 The established microwave antennas: horns and dishes

The paraboloid ('dish') secondary reflector, because it has the potential for a large aperture, has for long been one of the commonest microwave antennas. Paraboloids can be spun from metal (or turned in smaller sizes) at moderate cost, and because of their arch-like cross-section have good rigidity, even in rough weather. Again, the theory of these antennas is exactly as described in Chapter 6. At frequencies up to about 10 GHz it is even possible to use dipole feeds, often seen used with plate reflectors; however, with increasing frequency the feed is likely to be a waveguide horn, even though the waveguide run itself may be very short in modern equipment.

The **horn antenna** derives from a waveguide, and has long been an important microwave type in its own right, quite apart from its use as a paraboloid feed. In a waveguide the radio quanta pass along the axis of the guide, but as they do so the electromagnetic wave functions associated with them cause circulating currents in the walls of the guide. In consequence, if it is suddenly terminated the radio quanta do not proceed onward in an orderly beam, but instead suffer considerable deflection and reflection at the open end, due to the interruption of the pattern of current flow in the walls. The result is standing waves in the guide, with all the usual and undesirable resonance effects that causes, and a broad emitted lobe in the case of those quanta of radio energy that do escape. To overcome this problem the solution is similar to that adopted with a wire transmission line in the design of non-resonant antennas, as described in Section 7.3.

Provided the change in dimensions of the transmission line takes place slowly enough, ideally over many wavelengths, the reflection of energy can be made arbitrarily small. The same is true of waveguides. Sudden termination can be regarded as an instant increase of the cross-sectional dimensions to infinity, and causes severe energy reflection, but if the cross-sectional dimensions increase gradually the reflection can be reduced to an acceptable magnitude (Fig. 9.4). The aperture of such an antenna approximates very closely to the area of the open 'mouth' of the horn, and the flare length is adjusted to give an acceptably low level of standing waves in the waveguide. If the mouth area is A the gain of the antenna relative to isotropic is given by eqn (9.1) above. Although large horn antennas are occasionally seen, they are relatively expensive to construct and heavy, partly because of the close mechanical tolerances which must be held, while their only major advantage is the ability to handle very high power levels.

So for powers up to a few kilowatts and apertures exceeding a small fraction of a square metre it is a more economical solution to use a

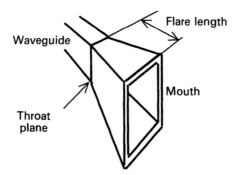

Fig. 9.4
A horn antenna.

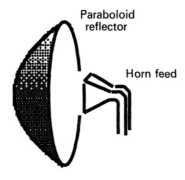

Fig. 9.5
A horn feed to a paraboloid secondary reflector.

horn antenna as a feed to illuminate a much larger secondary reflector, almost always a paraboloid. This is a very widely used configuration, seen everywhere functioning as a satellite television antenna, for example (Fig. 9.5). The characteristics of such antennas are little different from those of a paraboloid with a dipole array feed, and are described in Section 6.4. They offer a large aperture, high gain and a narrow main lobe at relatively modest cost. The principal difference between a paraboloid reflector for microwaves and one for lower frequencies is the closer tolerance with which the reflector as fabricated must follow the theoretical paraboloid shape.

146 Radio Antennas and Propagation

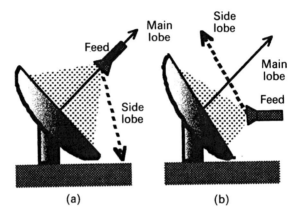

Fig. 9.6
An offset feed has advantages for the paraboloid antenna.

Simply placing the primary feed at the focus of the reflector 'dish' (Fig. 9.6(a)) has disadvantages. Along with its supporting members, it causes obscuration of part of the main lobe, absorbing radio quanta or scattering them in undesired directions, also the feed itself is somewhat inaccessible, and the feed side lobes 'see' the relatively noisy ground, which can be important when receiving very weak signals, as in satellite applications. By using an offset feed (Fig. 9.6(b)) these disadvantages are largely overcome, in particular most of the side lobe response points to the sky – cold and therefore less noisy.

Even better is the Cassegrain configuration (Fig. 9.7). Here the feed is moved behind the paraboloid, from where, through a small hole, it illuminates a convex reflector (of hyperbolic shape) which spreads the photons uniformly all over the paraboloid surface. This type of antenna is much more compact for a given focal length of paraboloid. Since long focal length paraboloids are desirable, having lower noise and being easier to manufacture, this is a substantial advantage. Also the Cassegrain antenna puts the feed in a very much more convenient place.

Offset forms of the Cassegrain antenna are sometimes seen, and in another variant the horn feed is replaced by a small active adaptive

Microwave antennas 147

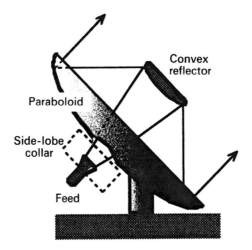

Fig. 9.7
The Cassegrain antenna uses an additional convex reflector.

array, which permits a degree of null and beam steering. This hybrid arrangement can give many advantages of the adaptive array, but with a larger aperture than would otherwise be economic. Sometimes a collar is placed around the feed, made of radio absorbent material (possibly cooled for low noise) which almost completely suppresses the side lobes. This could also be done with other configurations, but less conveniently. In the usual case of the very weak received signals, side lobe responses may provide access for co-channel interference, so it can be well worthwhile to reduce them. This is particularly important to the military, whose radar and communications are sometimes attacked by jamming into side lobes.

Paraboloid and horn antennas are sometimes required to rotate (for example, in radar systems where the rotation enables the beam to scan the target area). This is achieved by using a circular waveguide, which (with certain precautions at the join) can be cut at right angles to its axis to enable the two parts to rotate relative to each other. Rotating waveguide joints are commercially available which comprise a rectangular-to-circular waveguide transition, a rotating joint in the circular waveguide and then a circular-to-rectangular transition. These can be incorporated into conventional rectangular waveguide runs wherever rotation is required.

9.5 Refractive microwave antenna components

In his earliest experiments Heinrich Hertz managed to demonstrate the deflection of a beam of microwave energy by a wood prism. Material of high refractive index which is nevertheless transparent to microwave quanta can be used to deflect and focus beams just as glass components are in optical instruments. It turns out that a number of plastic materials have the desired properties and from time to time microwave antennas are constructed which exploit them. For example, it is perfectly possible to use a plastic plano-convex lens in front of a horn feed to produce a parallel beam of photons, instead of using a paraboloid reflector. The analogy is with refracting and reflecting telescopes in optics.

Such antennas have not been used much, primarily on grounds of cost. The apertures which are required, at least in the SHF band, make for very large lenses which are more costly than simple spun sheet-metal reflectors. An exception might be in very large production runs, which could absorb the high tooling costs of precision plastic injection moulding, but very few microwave systems are built in this volume. In the EHF bands reflectors themselves become more expensive to fabricate (because of the tighter dimensional tolerances) but this affects refractive components too, pushing tooling costs yet higher, and in addition the losses of plastics begin to rise, which makes them less attractive, enforcing the use of more exotic and expensive materials. Sometimes it can be economic to use refractive correction plates to introduce minor changes in the characteristics of reflector antennas, perhaps to offset the effects of low-cost manufacture of reflectors to simplified profiles. Because the effects of the correction plates are small it is acceptable to make them to loose specifications, and hence cheaply.

Fig. 9.8
A suitably formed plastic 'plug' in a waveguide end can substitute for a horn antenna.

It is also possible to terminate a waveguide in a plastic 'plug' which, if long enough and sufficiently gently profiled, prevents excessive reflection and produces a reasonably well-constrained lobe, similar to a horn (Fig. 9.8). This has application to the primary feed of a paraboloid reflector antenna and shows cost advantages on large enough volume of production. Several distinctive designs have been proposed, all of which were created by numerical solution of Maxwell's equations.

Problems

1. Why are antennas for higher frequencies often more directive than those for lower frequencies?

2. A satellite at 36 000 km from the Earth is required to cover an area of 600 km in diameter, but the antenna must fit into a space on the satellite 60 cm square. What type of antenna would you choose and what transmitting frequency? [paraboloid or an array, 29 GHz]

3. A point-to-point radio link has at one end a paraboloid antenna on a slender self-supporting mast 30 m high. In high winds the signal level is found to dip by 3 dB from time to time. Given that the antenna gain is 40 dB, estimate how far the top of the mast moves in the wind. [less than 47 cm]

PART TWO

PROPAGATION

Between the emission of radio photons by a transmitting antenna and their capture by a receiving antenna (or reflection by a radar target) they move from one location to the next with the speed of light. There are many different modes of propagation, and deciding which one should be adopted for the solution of any particular challenge to the radio engineer depends on matching its characteristics to the problems to be solved. Once the mode of propagation has been chosen it is generally straightforward to select the optimum radio band in which to work, and the best receiving and transmitting antennas. Identifying the manner of propagation is thus the first step in radio system design. The following chapters review the various processes of radio propagation.

We begin with propagation in free space, without atmosphere, the proximity of Earth or any intervening objects. This is of practical importance in communication between space vehicles, and is generally a good approximation to the radar case, but in addition forms the basis for understanding more complex situations. Next, the effects of the atmosphere are considered, both refraction and absorption, increasingly important above 30 GHz.

The effects of proximity of the Earth's surface are then taken into account. Surface waves can propagate over the Earth, and this is important in the MF and LF bands. At higher frequencies reflections from the Earth's surface and terrain features such as hills give rise to multipath propagation, which results in complicated interference phenomena. Diffraction effects also occur. In the

VHF bands and above other objects, such as buildings, vehicles and vegetation, have an increasing effect, by reflection, shadowing, diffraction or absorption. Many users wish to use equipment inside the places where they live and work, so propagation into buildings and from room to room within them is important.

Intercontinental radio communication has long been of great commercial importance. At VLF and LF surface wave propagation can be exploited, but intrinsic problems with these bands limit their general use. At higher frequencies, intercontinental transmission depends on directing radio energy away from the Earth's surface, to avoid excessive attenuation, and then in some way returning them again to Earth in the intended reception area. One approach to getting the radio quanta back from space exploits the trails of ionized gas left by meteors entering the Earth's atmosphere. Although most are no bigger than a grain of sand, in the thin atmosphere at high altitude even tiny meteors produce clouds of ionized gas many kilometres in dimension, so this improbable-seeming mechanism for long range communication proves reliable at ranges up to 2000 kilometres.

Since the early twentieth century, reflection of radio quanta in the HF bands by the ionospheric layers high above the Earth's surface has been used for radio communication, resulting in low-loss intercontinental propagation. Ionospheric propagation remains commercially important, boosted by the evolution of computer-controlled equipment, which can be used by unskilled operators.

Artificial satellites, however, are by far the most effective of all the long-distance radio communication technologies, supporting intercontinental radio communication with reliability, availability and channel capacity not feasible by any other means. At first geostationary satellites dominated, and they continue to be of the utmost importance, but subsequently the advantages of low Earth orbit satellites (LEOs) came to be seen, and highly significant systems, both civil and military, have been developed to exploit them.

CHAPTER 10

ELEMENTS OF PROPAGATION

Radio quanta, like all photons, have no electric charge and very little mass, so they move in straight lines in all but the very strongest gravitational fields, where their paths will be slightly curved by gravitational attraction. In fact the interaction of gravity with radio photons is such a weak effect that it can be neglected in all except the most rare situations encountered only in radio astronomy. We shall not consider it further here, and with that very slight proviso shall assume that radio quanta travel in straight lines and, of course, at the speed of light. If it is only the far field that is of interest they may be taken as originating from point sources since the transmitting antennas are far away. In idealized theoretical cases these are considered isotropic, that is they radiate photons uniformly in all directions, but in the real world they are non-uniform radiators, having a polar diagram which defines a more or less complex variation of density of emission of radio quanta with angle, ranging from the 'doughnut' shape of the dipole to the pencil beam of a paraboloid reflector. The first step is therefore to reconcile the complexities of actual antennas with the simple theoretical model.

10.1 Equivalent isotropic radiated power

In order to deal with the complexity of actual antenna radiation patterns, note that in any particular direction the density of radio quanta radiated is exactly the same as it would be if an isotrope

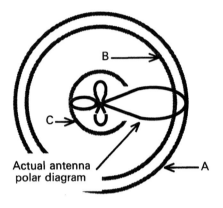

Equivalent isotropic polar diagram
A: for peak of main lobe
B: for main lobe half-power points
C: for peak of reverse lobe

Fig. 10.1
Equating an actual polar diagram to its isotopic equivalent.

were used, its power equal to the actual total power radiated by the real antenna multiplied by the gain of the antenna in the direction concerned (Fig. 10.1). (Although for convenience the polar diagram is shown in two dimensions, the argument which follows is equally valid in three.) In the direction of the peak of the main lobe an isotropic radiator with polar diagram A would produce an identical flow of radio quanta (but only in that one direction) to the antenna actually used. Radiation in the directions of the half-power points on the main lobe will correspond to that of the less powerful isotrope B, while in the reverse direction it would be the much less powerful isotropic radiator represented by C which equates to the actual antenna.

This leads to the notion of **EIRP** (equivalent isotropic radiated power). In any specified direction there is always an isotropic radiator which produces a radio quantum flow identical with that of the actual antenna, but which radiates a different total number of radio quanta integrated over the whole sphere, that is a different total power (since each individual photon carries a fixed quantum of energy dependent only on the radio frequency). This notional power radiated by the isotrope which produces, in a particular direction, the same radio quantum flux as the actual antenna is called the EIRP. Obviously, if the EIRP is P_I while P is the total power radiated by the actual antenna, then if G is the power gain

$$P_I(\phi, \theta) = G(\phi, \theta) \cdot P \tag{10.1}$$

In words, the EIRP in a certain direction is the actual power multiplied by the power gain in that direction. The EIRP concept is useful because using it we can develop propagation theory assuming the simple case of isotropic radiators, and then get the right answers simply by using the EIRP as the isotrope power.

Sometimes the EIRP is specified without giving the polar coordinates which indicate in which direction it is to be taken. In such cases it is the accepted convention that the direction intended is that of the peak of the main lobe. Sometimes this is called the **main lobe EIRP**.

10.2 Propagation in space

The simplest of all possible radio propagation environments is free space. We begin by assuming that neither atmosphere nor any solid objects are present to complicate matters. When transmitting, we will be concerned only with the far field, and are thus always very distant from the antenna, which can therefore be approximated as a point source, of negligible dimensions (Fig. 10.2).

Think of a small area ΔA distant r from an isotropic radiator at point T, which emits a total power P_I. The radio quanta are emitted uniformly in all directions, and therefore if the total number of photons emitted per second is N, each having energy hf, the number passing through ΔA is equal to ΔN per second where, from eqn (2.2),

$$\Delta N = \frac{N \cdot \Delta A}{4\pi r^2} = \frac{\Delta A}{4\pi r^2 hf} P_I \tag{10.2}$$

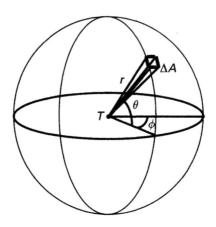

Fig. 10.2
Isotropic propagation in free space.

Note the inverse square law – just a consequence of the fact that in space the radio quanta do not get 'lost' by any mechanism and are therefore simply a constant number spread over a sphere which increases in area with the square of its radius. Sometimes, instead of the number of photons passing through this small area, we may prefer to see the power flow. This is the number of radio quanta per second multiplied by the energy of each radio quantum, and is thus

$$p = \frac{\Delta A}{4\pi r^2} P_I \tag{10.3}$$

If the radio transmission is received by an antenna of aperture (capture area) A_R the total power received is P_R where

$$P_R = \frac{A_R}{4\pi r^2} P_I \tag{10.4}$$

This simply says that the received power is the radiated equivalent isotropic power multiplied by the ratio of receiving antenna aperture to the total area of the sphere at that distance. This very well-known result appears in a variety of different disguises. For example, to get rid of the EIRP it may be re-written (using eqn

Elements of propagation

(10.1)) as

$$P_R = \frac{A_R G_T}{4\pi r^2} P_T \qquad (10.5)$$

where G_T is the transmitting antenna gain, P_T the transmitted power.

This is a very useful form of the expression, since in transmission the effect of antenna directivity is best described by the power gain, whilst when receiving it is the aperture which is more significant. Aperture and gain are related, however, so sometimes the expression is re-written in terms of the receiving antenna gain, when it becomes downright deceptive, hiding what is going on. From eqn (3.5), above

$$A_R = \frac{\lambda^2 G_R}{4\pi}$$

In consequence eqn (10.5) can take the form (often seen)

$$P_R = \frac{G_R G_T \lambda^2}{16\pi^2 r^2} P_T$$

This variant of the expression might give unwary readers the impression that somehow radio propagation in space is wavelength dependent. It most certainly is not. Radio quanta travel through free space in just the same way regardless of their energy, and hence wavelength. The aperture of the receiving antenna is the thing that varies with wavelength (Fig. 10.3) assuming constant antenna gain. Antennas like paraboloid reflectors which have a nearly constant aperture equal to their physical area also have a gain which is inversely dependent on the square of wavelength, so here too the wavelength term in the expression for received power gives a false impression.

Using eqn (10.5) or one of its variants it becomes easily possible to calculate the received power over a link in free space between two antennas of specified type.

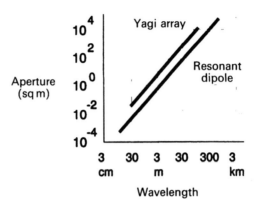

Fig. 10.3
Variation of aperture with wavelength.

10.3 The case of radar

Radar systems are used to locate a target object in space; they generally involve transmissions which do not interact much with the Earth's surface or objects other than the target, and are therefore reasonably well approximated by free-space propagation. In the simplest form, position of the target is obtained in polar form (r, θ, ϕ) from a single site where transmitter and receiver are collocated.

Angular position is most commonly derived by using an antenna with a narrow main lobe, producing a 'pencil beam', common to reception and transmission, so that the angular orientation of the main lobe of the antenna corresponds to the direction of the target. The obvious antenna design for this service is the paraboloid, which was used very extensively in the early years. Antennas were mechanically scanned over the area to be surveyed and the operator's display was synchronized with the movement of the antenna so that the angular location of the origin of received signals was directly displayed. More modern radars use active adaptive arrays with very fast electronic steering of the main lobes. Direct digital readout of the orientation of lobes contacting targets is immediately available. A smaller array as a feed to a paraboloid can also sometimes be seen, often scanned in the vertical direction only by frequency variation, the horizontal scanning being mechanical. The falling cost of large digitally controlled active arrays gives this compromise a limited future.

Range information is obtained by measuring the 'round trip' time for quanta emitted from the antenna, reaching the target and returning to the radar receiver. Systems are of two types: **primary radar** and **secondary radar**. In the case of the commercially very important secondary radar, the target receives a signal from the radar transmitter on its own antenna and carries a **transponder** which amplifies and then re-transmits it, often on another channel. In this case we can consider the radar propagation as simply equivalent to two point-to-point radio channels end to end. Equation (10.5) can be used to compute the signal received by the target's antenna, and then again to compute the return signal, knowing the transponder output power. The advantage of secondary radar is that the transponder can add useful information to the return signal, such as the target identity, and the returned power can be high regardless of the received signal level, depending only on the transponder design. For this reason air traffic control uses secondary radar wherever possible. However, there will be times when aircraft transponders are defective or they are not fitted, so co-operation by the target cannot be assumed in all cases. Thus there must always be strong pressure to have primary radar available, at least as a last resort, since it relies simply on radio energy passively scattered back from the target.

To measure the 'round trip' time it is necessary that there should be some identifiable marker on the transmitted signal, and in practice this will take the form of some kind of modulation. The early radars invariably used pulse modulation, primarily because the radio frequency power devices then available, such as magnetrons, were more suited to pulse generation. The time for a pulse – in effect a packet of quanta – to travel to the target and return was used to measure the range. This process has many disadvantages, but particularly arising from the very wide Fourier spectrum of pulses, which meant that receivers had to have wide bandwidths and were therefore very susceptible both to noise and jamming.

More recently there has been a general move to **continuous wave** (CW) radar, in which the carrier is not pulsed but instead is modulated in some other way. The simplest possibility is sinusoidal modulation, in which case the time taken for a signal return can be obtained by measuring the phase shift between the modulation on outgoing and returning signals. Because there is the possibility of ambiguity if the shift comprises several cycles of modulation, yet a short modulation period is required for good range resolution, it is not uncommon to have more than one sine wave, of very different frequencies, the lower one to resolve ambiguities and the higher to give low range error. Another approach is to modulate the signal with a long binary sequence, in which case the time delay between outgoing and incoming sequences can be determined by a process of correlation. This wholly digital approach is often instrumentally rather simple, and it gives resolution to better than one bit duration whilst ambiguity is impossible provided that the whole sequence is longer than the longest possible 'round trip'.

Finally, in regard to ranging, note that whereas for primary radar it is reasonable to assume that the process of reflection takes place instantaneously, in the case of secondary radars it is important that the time delay through the transponder should either be negligible or at least known and stable in order that the correct 'round trip' time can be obtained.

10.4 The primary radar power budget

We consider the case of a primary radar. When radio quanta are incident on a target they will be scattered in every direction to some extent, but a proportion will return in the direction from which they came. Considering the EIRP in the back direction, P_B, if the incident power per unit area is p (eqn (10.4)) then

$$P_B = pA_C$$

where A_C is a constant.

The constant A_C, which has the dimensions of an area, is known as the **radar cross-section** of the target. This depends on the size and character of the target, and can vary from less than 1 sq m, for a person or a small pilotless drone aircraft, to more than 10 000 sq m for a ship. Using eqn (10.5), the signal power returned to the radar receiver is

$$P_R = \frac{A_R}{4\pi r^2} P_B = \frac{A_R \cdot A_C}{4\pi r^2} p$$

Substituting a value for p from eqn (10.3)

$$P_R = \frac{A_R \cdot G_T P_T}{16\pi^2 r^4} A_C \tag{10.6}$$

This is the well-known **radar equation**, almost comparable in importance to eqn (10.5) and like it appearing in a variety of essentially interchangeable forms. In particular, if the same antenna is used for transmitting and receiving, then

$$A_R = G_T \frac{\lambda^2}{4\pi}$$

so

$$P_R = \frac{G_T^2 \lambda^2 P_T}{64\pi^3 r^4} A_C \tag{10.7}$$

This is another useful, though less general, form of the equation. In both, the received energy varies as the inverse fourth power of distance from the target in this primary radar case, compared with the inverse square law for a secondary radar or free-space communications link. This limits the range, but even so modern primary air-defence radars can have ranges of as much as 500 km. Obviously the radar cross-section of the target is critical. If all other factors are kept constant, eqn (10.6) indicates that the range at which a target can be detected varies as the fourth root of the cross-section.

Because they wish to frustrate radar detection by the enemy, military designers have expended much effort on reducing radar cross-section in ships, aircraft and tanks. Vehicles with reduced radar cross-section are often known as '**stealth**' vehicles. Apart from simply minimizing the size, this can be achieved in two ways. First, careful attention is given to the structure of the vehicle, and in particular to avoidance of extended flat surfaces or of sharp included angles, which trap quanta and scatter them back. Second, the reflectivity of the structure may be reduced, either by covering it with radar absorbent material. This is believed to be in the form of a lossy plastic coating containing conducting components, but the exact details are military and commercial secrets. They virtually eliminate radar reflections over their designed range of frequencies.

The value of the radar cross-section is dependent on the aspect from which the target is illuminated by the radar transmitter (for example, an aircraft or ship seen approaching head-on will generally have a much lower radar cross-section than one seen broadside). Trying to reduce the effects of stealth design of targets, some air-defence radars are constructed in **multistatic** form (Fig. 10.4) or **bistatic** if there are only two sites. Instead of the radar transmitter and receiver being co-sited in the usual way (a **monostatic** radar (Fig. 10.5)), the transmitter (sometimes referred to as the **illuminator**) may be sited well away from the receiver. Indeed one illuminator may serve several receivers. The receiving sites must know the timing of the transmitted modulation from the illuminator with very good accuracy, and this can be done either by directly receiving illuminator transmissions or by linking the sites by radio point–point links or optical fibre. Because the reflection may occur on almost any part of the target in a multistatic system, it becomes necessary to extend low cross-section design techniques over the whole of the aircraft, which increases the magnitude of the task for the aircraft designer.

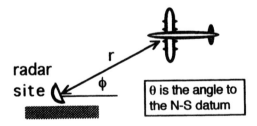

Fig. 10.4
A multistatic radar.

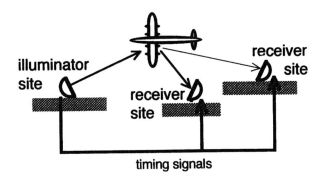

Fig. 10.5
A monostatic radar. θ is the angle to the N–S datum.

An intriguing possibility is to use as the illuminator (which does not need to have a narrow beam or to scan) some transmission which already exists for an entirely different purpose, such as television broadcasting. Since the receiving sites are entirely passive they would be invulnerable to attack by radar homing missiles, and the target would be entirely unaware that it was being detected. It is impossible to say whether such systems exist; if they did they would surely be highly classified.

Although first developed against military requirements, multistatic radar systems may have advantages for civil users also. A single illuminator can cover a wide area serving several receiving sites (at which the angular and range measurements take place), so the system is attractive for wide area coverage with minimum spectrum occupancy.

Problems

1. A transmitter in space radiates 10.0 mW at a frequency of 1.52 GHz. Its antenna has a gain of 20 dB relative to isotropic in the direction of transmission. How many quanta will be received per second by a parabolic antenna of area 1 sq m situated at a distance of 100 000 km? [8 million]

2. A primary radar radiates 100 kW (+80 dBm) and has a transmitting antenna gain of 40 dB. The receiver requires a minimum received signal power of −113 dBm at the antenna to give an acceptable assurance of target detection and the antenna aperture when receiving is 50 sq m. If the same antenna is used for transmission and reception, what is the frequency of the transmissions? [1.2 GHz] What is the minimum radar cross-section that a target must have in order to be detected at 280 km? [1 sq m] If a transmitter failure reduces the power to 1 kW, at what range will this target now be detected? [157 km]

3. A target aircraft has a radar cross-section of 5 sq m and when approaching at maximum speed is detected by a primary radar 120 s before its arrival overhead. If a forward-firing missile defence battery near the radar is known to take 30 s to respond, to what must the aircraft cross-section be reduced in order that it shall be safe? [0.019 sq m]

4. A television transmitter at 846 MHz has an EIRP of 1600 MW in the horizontal plane but 40 dB less in a vertical direction. Its transmissions are used as an illuminator by a covert bistatic radar receiver which will respond to −120 dBm from its antenna, gain 10 dB. The target is high altitude aircraft of radar cross-section 1.4 sq m at this frequency. Given that the receiving site is adjacent to the television transmitter, what is the maximum altitude at which a target will be detected? [20 km or about 66 000 ft]

5. A helicopter is fitted with a 95 GHz downward-looking radar used to image the landing site to ensure that it is free of obstructions. Scanning array antennas with 20 dB gain are used for both transmission and reception, the transmitter power is 1 W and the receiver will respond to −70 dBm at the antenna. What range do you expect, neglecting atmospheric absorption and assuming that the ground reflects 1% of the quanta incident on it? [125 m] (Hint: If the ground is a mirror, what we have is like a point–point link over a distance equal to twice the range from helicopter to ground, but with additional attenuation due to imperfect reflection.)

CHAPTER 11

THE ATMOSPHERE

If an atmosphere is present between the transmitting and receiving antennas it is still usual to speak of **space wave propagation** provided that the proximity of the Earth or other material objects does not affect the propagation. Before looking in detail at its effects on radio transmission some of the general properties of the atmosphere will be summarized briefly.

The **troposphere**, the lowest layer of the Earth's atmosphere, has the **tropopause** as its upper boundary and the Earth's surface for its lower boundary. Here, significant flow of heat and water vapour occurs (usually upward during the day and downward during the night). Virtually all weather, or short-term variation in the atmosphere, occurs in the troposphere, where the air is principally composed of nitrogen (78%), oxygen (21%), argon (0.9%), carbon dioxide (0.03%) and water vapour (highly variable). The density and pressure of the air decreases by a factor of two for every 5.6 km increase in altitude above the Earth's surface. As a consequence the troposphere contains 99% of the atmospheric water vapour and 90% of the air. Tropospheric air temperature decreases with increasing height, except rarely when **inversions** occur.

The tropopause is the boundary between the troposphere and **stratosphere**. Because it is far from the Earth's surface and the ozone layer, two good absorbers of energy from the Sun, the tropopause is colder than the regions above and below it.

Above the tropopause, about 15 km in the tropics and 10 km near the poles, is the stratosphere, where the temperature no longer decreases with altitude. It extends from the tropopause to the **stratopause** (the 'ozone layer'). Ultraviolet absorption by ozone causes the high temperatures of the stratopause, which is usually about 50 km above the Earth's surface. Because the stratopause is relatively hot, temperatures in the stratosphere increase with increasing height. The stratospheric air flow is mainly horizontal. Above the stratopause are the **mesosphere**, which extends up to about 80 km, the **thermosphere** which goes to some 640 km, and finally the **exosphere**.

However, from the radio engineer's point of view, of particular significance is the **ionosphere**, from 80 to about 600 km altitude and overlapping the mesosphere and thermosphere. This is the part of the Earth's upper atmosphere where ions and electrons are present in quantities sufficient to affect the passage of radio quanta. Such a region was first identified in 1902, independently by the British physicist Oliver Heaviside (1850–1925) and the American electrical engineer Arthur Edwin Kennelly (1861–1949), following Guglielmo Marconi's radio transmission across the Atlantic on 12 December 1901. It makes possible intercontinental radio communications in certain bands, and will be considered in detail in Chapter 13.

We begin, however, by looking at radio propagation in the troposphere, which is far and away the most significant, commercially and socially. This, after all, is where we live.

11.1 Effects of an atmosphere on radio propagation

For radio transmissions within the atmosphere two important effects may occur, namely refraction and absorption. Refraction, which we will consider first, is bending of the path taken by radio photons, and occurs because the speed of a radio quantum which travels surrounded by matter, even a gas, is less than it would be in free space, due to a retarding effect by the surrounding atoms. It is precisely refraction, but applied to light photons, which is exploited to make lenses, prisms and the familiar components of optical systems and instruments. Similar things can be done at radio

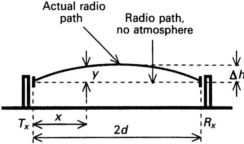

Fig. 11.1
The refracted radio path relative to the Earth's surface (curvature exaggerated for clarity).

frequencies; in his 1887 experiments Hertz used a prism made of wood to deflect the path of a radio transmission. To this day plastic lenses are occasionally used with microwave antennas to collimate the radio beam.

The **refractive index** of a transmission medium, such as air, is the ratio of the speed of photons in space (c) to the lower value it has in the material medium; this index is greater than one and for air is dependent on density (and pressure). The Earth's atmosphere has a refractive index largest near the surface and falling with height. It acts somewhat like a prism, tending to curve the path of radio quanta down toward the surface This has traditionally been analysed as equivalent to photons taking a straight path over an Earth of reduced curvature (increased radius) by a factor close to 4/3.

Communication between two points (Fig. 11.1) is over a path curved by refraction. The path can be approximated by a second-order equation, so

$$y = \Delta h \frac{x}{d}\left(2 - \frac{x}{d}\right) \qquad (11.1)$$

But what is the value of Δh? One would expect Δh to be proportional to d (which sets the scale) and also to d/a, a measure of curvature of the surface.

Detailed consideration leads to

$$\Delta h = k\frac{d^2}{a}$$

Here, a is the Earth's radius, and k is a constant $= 0.25$.

Thus

$$y = 0.25d\frac{x}{a}\left(2 - \frac{x}{d}\right) \qquad (11.2)$$

Practically, this refractive bending of the radio path is very significant. Consider the (fictional) contour map of Fig. 11.2. It is required to establish a VHF radio link between Newtown and Winterbourne, using an existing mast at Newtown which places the antenna 50 m above local ground level. What height of mast will be required at Winterbourne?

To solve a problem of this kind the contour map is used to prepare a section over the proposed path, by plotting elevations against distance as shown. Obviously the problem with this transmission

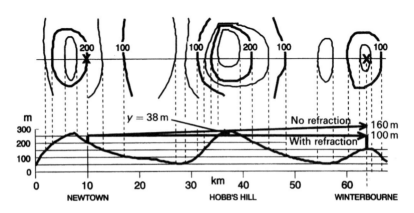

Fig. 11.2
Using a map section to determine an antenna mast height.

path is Hobb's Hill, which rises to over 250 m height between the two radio sites and 28 km from Newtown. Drawing in a straight radio path which just clears the hilltop suggests that a mast 160 m high will be required at Winterbourne; however, due to the effect of refraction the radio path will, in fact, be curved and therefore a lower nominal line is possible.

Using eqn (11.2) and noting that d is 27 km, at $x = 28$ km, the upward shift of the path is given by $y = 29$ m. We can therefore set off a straight line for the radio path which falls far below the summit of the hill, knowing that it will still just clear. This leads to a mast height at Winterbourne of 100 m. Compared with a 160 m mast, the cost will probably be less than half and the environmental impact will be appreciably improved.

Although eqn (11.2) could be used to calculate the whole radio transmission trajectory, in practice this is unnecessary as it is almost always sufficient to ensure clearance at one or two potential blockage points, as in the above example. Of course, when a radio transmission path grazes the top of a hill or ridge, strictly speaking this is no longer space wave propagation, and diffraction effects occur, as described in Chapter 12. It is also necessary to give consideration to what is actually to be found on the hilltop; if it were thickly wooded this could give rise to quite serious additional attenuation. Further discussion of these points will be postponed until later.

Even where hills are not a problem, over sea or very flat terrain, refraction effects enable radio paths to follow the curvature of the Earth's surface a little better, so that the 'radio horizon' is further than the visual one. Remembering that the 'hill' due to curvature of the Earth has a height midway between two sites on notionally flat terrain equal to d^2/a, in the absence of refraction this would be the minimum acceptable mast height h_m, giving a maximum range between similar masts of $2d$ where d^2 is ah_m. Using eqn (11.2) to calculate y for $x = d$, it is easy to show that the half range with refraction, say D, is given by $D^2 = 1.33 ah_m$ (note the 'four-thirds Earth' radius) so that the effective range is increased by about 15%. Alternatively, for the same range, masts can be used which are 25% shorter, with substantial savings in cost.

Evidently, when designing point-to-point links, refraction is a useful property of the atmosphere, from the radio engineer's point of view. The coverage area of broadcasting stations is similarly enhanced. However, this analysis assumes that the air is uniform in its properties, which thus vary significantly only with altitude, due to the change in pressure and temperature. Sadly, this is not always the case and less benign effects also occur.

Atmospheric refraction can introduce a small error into radar systems, particularly when the target is near the horizon. The elevation angel is overestimated. This effect is usually not big enough to be of any real significance in civil systems but can be a source of error in anti-aircraft weapon targeting systems.

11.2 Ducting

Although in the troposphere both the air temperature and pressure fall steadily as one goes higher for the most part, anomalies do occur associated with weather. Significant changes in pressure, temperature and water vapour content in large volumes of air are commonplace. Sometimes, in particular meteorological conditions, a layer forms in the troposphere with abnormal refractive index due to a **temperature inversion** (Fig. 11.3).

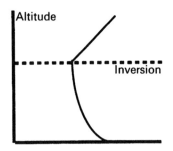

Fig. 11.3
The effect of temperature inversion on the refractive index of the atmosphere.

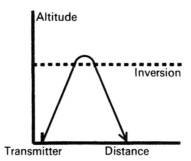

Fig. 11.4
A sky wave is returned to Earth by an inversion in this surface duct.

An inversion occurs whenever the normal decrease of atmospheric temperature or water vapour with height is reversed over a short vertical distance to produce a layer in the atmosphere where the temperature actually increases with height. Strong inversions form in the middle and upper latitudes in stagnant high-pressure areas. A stable inversion can form when cold air blows over warmer ocean water; in the tropics this is called the **trade-wind inversion**. Alternatively, a layer of cold air may form on top of the highly turbulent boundary layer over land in hot weather. Sometimes a **frontal** inversion occurs when warmer air flows over a mass of cold air.

If strong enough, the inversion layer is capable of reflecting radio waves, particularly in the VHF and the lower UHF bands (but occasionally at the high end of the HF band) (Fig. 11.4). When this happens a **surface duct** is said to have formed. This results in anomalously long-range propagation by means of a **sky wave**. Both surface and **elevated** ducts are seen; an elevated duct forms between two strong inversion layers one above the other, in which case the result can be that radio quanta are trapped between them at high altitude and propagate with low loss until the duct peters out, at which point they return to Earth. There have even been occasional reports of surface ducts forming in the tropics below the tops of exceptionally high antenna masts, so that the transmissions are trapped above the duct and virtually no radio energy reaches ground level.

Almost all VHF/UHF terrestrial propagation over long distances is attributable to ducting. In western Europe ducting is far too dependent on relatively rare meteorological conditions for it to be the basis of any useful service; it is simply an erratic source of interference and hence a nuisance (except perhaps to radio amateurs, who use ducting to achieve record-breaking long-distance transmissions). Ducting is the cause of summer interference with television, mobile radio and other services in the VHF and low UHF bands. Because it is worst in the VHF band, it has been a cause of a decline in use of this band in Europe for those services, such as television, which are particularly susceptible to interference. In other geographical areas, however, things are very different. Due to the high temperature and humidity, there is a near-permanent strong duct over the whole Red Sea and Gulf area, with the result that TV transmissions are routinely received at distances up to 300 km, providing a useful service in areas that would otherwise not be covered.

11.3 Atmospheric absorption

As well as causing refraction, the presence of the atmosphere can also lead to **absorption**, although the effect is important only in the EHF band (>30 GHz) where it leads to so-called **absorption bands** and **windows** both of which have their uses. Absorption by the atmosphere is due to collisions between radio quanta and gas molecules, in which the photons give up their energy. This is only highly probable if the gas molecules have the possibility of changing their quantum energy state in such a way that they can absorb nearly the exact energy carried by the radio quantum, equal to hf. Thus the **cross-section for collision** of the molecules is highly frequency dependent. The quantum states of molecules are fixed by their internal structure, and are thus different for each different molecule. As a result, each type of matter through which radio quanta pass, such as the gases of the atmosphere, has its own distinctive **absorption spectrum** (Fig. 11.5), characterized by absorption peaks where the frequencies are just right to hit one of the maxima of collision cross-section.

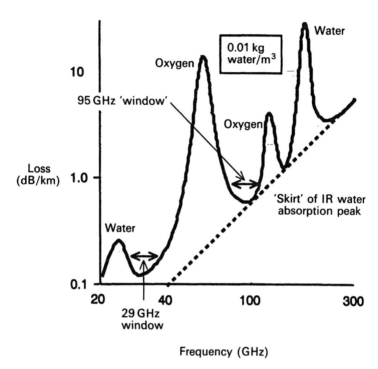

Fig. 11.5
Absorption by the atmosphere. (Source: van Vleck, J.H. *Phys. Rev.* **71** (1947).)

Several features of this graph are interesting. A large absorption peak occurs at 60 GHz and another at 120 GHz; these are due to collisions of radio photons with oxygen molecules. These peaks do not vary in size, independent of geographical location or weather conditions, because the amount of oxygen in the atmosphere is virtually constant. This is not true of other features of the curve. The peaks, at a little over 20 GHz and another at just under 200 GHz, are due to water vapour, as is the rising 'floor' of the absorption curve, which in fact is the skirt of a very large peak in the infrared optical region. All of these features are a consequence of the water vapour content in the air, here assumed to be about the value for a dry day in western Europe. However, the water vapour content is heavily dependent on both the temperature and the weather conditions. It will be far more in a hot, wet tropical environment and far less in an arid desert. In consequence, both the 'floor' of the absorption curve and the water peaks will be very dependent on weather and location.

The 60 GHz absorption band has been widely advocated as a way of improving the ratio of service to interference range for a radio transmitter. Because the signal falls at 14 dB per/km in addition to the usual inverse square effect, it rapidly becomes negligible beyond the service range. The signal loss due to increasing range, expressed in decibels, can be written as the sum of two components

$$L = L_{\text{spread}} + L_{\text{absorption}}$$

The spreading loss term can be obtained from eqn (10.4) earlier, hence we may write

$$L = 10\log_{10}\left(\frac{A_r}{4\pi r^2}\right) + \beta r \tag{11.3}$$

Here, β is the absorption coefficient, and has the value -14 dB/km (-0.014 dB/m) at the first oxygen absorption peak. This figure, although high, is constant, so provided that the operating range of a link is fairly short it can be engineered at this frequency to operate quite reliably. However, the EIRP would have to be increased by a little over 14 dB (nearly 30 times) for every additional kilometre of range required. The signal range is thus well defined by the oxygen absorption. Probably the main commercial interest in the 60 GHz band is for short links where the attenuation can be tolerated and the negligible remote interference capability (either from the main or side lobes of the antenna) means very wide bands of frequencies can be assigned in particular locations.

Military users have been interested in the 60 GHz absorption band as a means of radio communication which is difficult to intercept at a distance. The detection and identification of radio signals, known to the military as **ESM** (electronic support measures), is an important part of electronic warfare; they further subdivide it into **elint** (identification of signals by their technical characteristics, such as waveform and spectrum) and **sigint** (identification of the information carried by signals). Both are made more difficult the weaker the signal being intercepted, and therefore are possible in the absorption band only at very short ranges.

Also of interest are the so-called 'windows', relatively low absorption regions in the EHF band. However, such windows have a 'floor' attenuation determined by residual water vapour absorption, and therefore are strongly dependent on weather conditions. The 29 GHz window extends from about 29 to 38 GHz and is used for both point–point communication and radar. As links are rarely more than 10 km long, the attenuation of 0.2 dB/km in the window will contribute up to 2 dB of excess loss, which is acceptable. However, this is only true in dry conditions, and things get measurably worse in rain of more than 0.5 mm/h. In significant rainfall the attenuation increases sharply, roughly proportionally to the rate of precipitation. At 5 mm/h (light rain) the excess attenuation will be 2 dB/km. Another significant window occurs at 95 GHz, although the attenuation here is four or five times larger (in decibels), still further reducing the range, which in this case also is very weather dependent. The 95 GHz window is primarily used for short-range precision radar. However, note that the atmosphere rapidly thins with altitude (half is gone in 5.6 km) so the absorption also falls quite fast. For this reason the use of the EHF band is quite practicable for satellite communication, and is increasingly seen as attractive. We shall return to this in Chapter 13.

The principal attraction of the EHF band for radar designers is that it is possible to construct highly directional (and high-gain) antennas which are physically not very large. Thus, target location can be very accurate (because of the small angle of the antenna lobe); however, for the reasons indicated above the attenuation will be severe, particularly in rain, so the range is short. These frequencies are therefore used for precision imaging radars of 1–2 km range, such as for systems that detect people, for aircraft landing aids or, among the military, for tank weapon targeting. This last is important because of the high dust levels in most tank battles, which make visual targeting difficult.

Problems

1. The two ends of a radio link are 100 km apart. How far is the radio path above the direct line 25 km from the transmitter? [73 m]

2. An ESM station receives VHF transmissions from an aircraft approaching at an altitude of 1000 m flying at Mach 2.5. How much sooner will it do so as a result of atmospheric refraction than if this were not present? [16 s] A strong surface duct forms over the whole area at a height of 500 m. What will be the effect on detection time? [no detection] (Assume Mach 1 = 300 m/s.)

3. A 60 GHz transmitter of power 0.1 W has a transmitting antenna of 40 dB gain. At a distance of 5 km the receiving antenna has an aperture of 0.2 sq m. What is the received power? [−102 dBm] If the distance were increased 2 km by what factor would the loss increase? [31 dB]

4. A 10 km point-to-point link at 30 GHz uses antennas at each end of 30 dB gain and requires a received power of −80 dBm for satisfactory operation. What transmitter power must be used if the system is to continue to function in rain at 5 mm/h? [64 mW or +18 dBm] (Take absorption as 1.6 dB/km.)

CHAPTER 12

AT GROUND LEVEL

Terrestrial communication at or near ground level is of the greatest commercial significance because human populations are concentrated there. The previous chapter considered the effects of the atmosphere on the passage of radio photons, but even more important is the influence of solid objects, the largest of which is the Earth itself. These lead to reflection, shadowing and diffraction effects which greatly complicate the prediction of received radio signals. We begin with the simplest case of communication over the Earth's surface and see how it is influenced by reflection.

12.1 Reflection

In Chapter 6 we have already considered reflection of radio quanta in connection with plate antennas and those using secondary reflectors. In fact the near-earth terrestrial environment is usually full of reflectors. From VHF to microwave, most building materials are partial reflectors of radio quanta and metals almost wholly so. Materials which reflect poorly, such as glass, are often backed up (in glass clad buildings) with other materials (metal, building blocks) which reflect better. Thus virtually all normal buildings and almost all road vehicles can be treated as good reflectors of radio energy in these bands. So also can hills, mountains and other terrain features. Above all, the surface of the Earth itself reflects, and at all frequencies down to the lowest, due to its large size.

178 Radio Antennas and Propagation

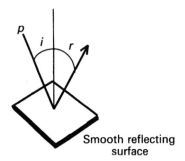

Fig. 12.1
Specular reflection.

Classically, two kinds of reflection can be distinguished: specular and diffuse. **Specular reflection** (Fig. 12.1) is the case where the surface can be treated as perfectly smooth at the frequency of interest, meaning that any surface irregularities are very small compared with a wavelength. The angle of incidence i is then equal to the angle of reflection r and if p is the incident power density (W/sq m) the total reflected power density p_r is

$$p_r = Rp \qquad (12.1)$$

where R is the reflection coefficient, a positive number always less than one, but close to it in favourable cases. Note that in determining whether a surface is smooth enough for reflection to be specular, wavelength is all important. The side of a hill may be quite smooth on the scale of 100 m-wavelength (frequency = 3 MHz), but quite rough at 300 MHz where the wavelength is only 1 m. However, a stone wall may look mirror-like at 300 MHz yet very rough indeed at 30 GHz (the wavelength being 10 mm).

The second type of reflection is **diffuse reflection**, which occurs mainly at shorter radio wavelengths, where objects (walls and so on) can be considered as very rough on the scale of the wavelength. In this case Lambert's cosine law applies

$$p_r(r) = R' \cdot p \cos(i) \cdot \cos(r) \qquad (12.2)$$

and there is some energy scattered over the whole hemisphere.

So far as radio propagation is concerned, the most dramatic effects are caused by specular reflection, which is very common. For photons to be reflected the objects acting as 'mirrors' will generally have dimensions of the order of at least a wavelength and usually much larger, although strong reflections can occur from smaller objects which happen to be near resonant dimensions (c.f. parasitic reflectors in Yagi arrays). However, resonant reflectors are rare at frequencies in the MF and lower, due to the scarcity of large enough conducting objects.

The commonest non-resonant reflectors in the radio bands can be summarized as:

ELF, VLF	Earth's surface (but deep penetration), ionosphere
LF, MF, HF	Earth's surface, ionosphere
VHF, UHF	Earth's surface, mountains, buildings, vehicles, ducting and ionospheric effects, meteor trails
SHF	Earth's surface, mountains, buildings, vehicles, trees, people, street furniture
EHF	Just about everything, excluding upper atmosphere effects

Changes from band to band are not as sharp as this list suggests, instead there is considerable overlap at transitions.

12.2 Multipath phenomena

As an example of the effects of radio reflection by the environment, consider the case of two hilltop radio stations which communicate over a flat plain. There are two paths taken by the radio quanta travelling from the transmitter to the receiver: the direct path, and a ground-reflected route. This is the simplest example of **multipath propagation**, one of the most important phenomena of radio systems, producing a variety of very significant effects.

The direct and reflected paths are not of the same length, so photons travelling by the two paths may arrive with their associated wave functions in virtually any relative phase. The difference in length is

$$\Delta = 2\sqrt{h^2 + d^2} - 2d$$

or, since h is always very small compared with d, using a binomial expansion

$$\Delta = 2d\left(1 + \frac{h^2}{2d^2}\right) - 2d = \frac{h^2}{d}$$

which produces a phase shift between the two wave functions arriving at the receiving antenna equal to

$$\phi = \pi + \frac{2\pi}{\lambda} \cdot \frac{h^2}{d} \quad (12.3)$$

The first π on the RHS of this expression is due to the phase inversion on reflection, as described in Chapter 6.

For

$$h = \sqrt{nd\lambda} \quad \text{(where } n = 0, 1, 2, 3\ldots\text{)} \quad (12.4)$$

the received waves are in antiphase, and hence sum to minima, corresponding to a minimum in the number of radio quanta, and hence the power density.

For

$$h = \sqrt{md\lambda} \quad \text{(where } m = n + \tfrac{1}{2}\text{)} \quad (12.5)$$

the received waves are in phase, and therefore sum to maxima. Thus, perhaps surprisingly, with a ground reflection the height of the antenna can be critical to received signal strength, even where there is no question of obstruction of the line of sight. This sensitivity to small changes in antenna height is due to **interference**

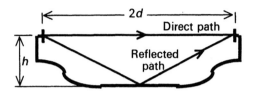

Fig. 12.2
Transmission over two paths, direct and reflected.

between two wave functions, which sum to maxima or minima depending on their phase relationship. Interference phenomena dominate terrestrial radio engineering.

In the case described (Fig. 12.2), the two paths (direct and reflected) are one above another, in the same vertical plane, and they introduce unwanted antenna height sensitivity. Had they been in the same horizontal plane, with the reflection off the side of a large building or a mountain, for example, they would have resulted in sensitivity to horizontal displacement. This too is very often seen.

Interference produces more than just sensitivity to the spatial location of the receiving antenna, however. Because the wavelength appears in eqn (12.4) the effects of interference are frequency dependent. The equation can be re-written as a condition on the frequency for minima as

$$f_{min} = \frac{ndc}{h^2}$$

Similarly the frequency for maxima is given by

$$f_{max} = \frac{mdc}{h^2} = \frac{(n+\tfrac{1}{2})dc}{h^2}$$

So the difference Δf between adjacent (same n) maxima and minima frequencies is

$$\Delta f = \frac{dc}{2h^2} = \frac{c}{2\Delta} \qquad (12.4)$$

Δf is called the **coherence bandwidth**. Difficulties will be encountered if the modulation bandwidth approaches the coherence bandwidth.

Multipath effects are likely to occur in all radio systems operating other than in free space. Multipath phenomena have two marked effects:

1. Variation in the level of received RF due to the phase relationships between the received signals. This is sometimes called **fading**, and is particularly noticeable in mobile systems where the receiving antenna may pass through maxima and minima as it moves.

2. Confusion between modulation carried by the signals arriving over the shortest paths and those which are more delayed. This could produce slight echo effects on analogue voice transmissions but, much more importantly, can also give rise to confusion between successive digits in a digital transmission, particularly at high bit rates. If the bit rate is B bits/s, evidently errors are probable for all path length differences $\Delta \sim c/B$ or greater. Thus if the bit rate B is n Mbits/s the critical path difference is $300/n$ m, which can easily arise for $n > 1$.

12.3 Shadowing and diffraction

Solid objects in the radio propagation environment not only reflect radio quanta but also obstruct and absorb them, giving rise to **radio shadows** (Fig. 12.3). However, the region of the geometrical shadow is not entirely free of photons; at the edge of the obstruction an effect occurs which results in some energy being thrown into the shadow region. This effect is known as **diffraction**. Its theory was first worked out for the optical case by Augustin Fresnel. His ideas apply with little modification to radio propagation.

At ground level 183

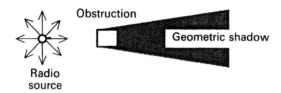

Fig. 12.3
Solid objects can absorb or reflect quanta to create radio 'shadows'.

> The French physicist Augustin Fresnel (1788–1827) advanced both theoretical and applied optics. Following in the footsteps of Christiaan Huygens (1629–1695) and Thomas Young (1773–1829), Fresnel rejected Newton's idea that light consists of particles and instead championed a wave theory. It lasted for a century, until the coming of quantum mechanics. An important part of his advocacy of the wave theory was his explanation in 1814 of the phenomenon of diffraction as applied to light.

There are two important ways in which light is diffracted. If it passes through a very small hole it does not form a sharp image on a screen but instead a series of bright and dark rings, or fringes, are formed, some of which fall within the predicted geometric shadow. The second type of diffraction occurs when light falls on an opaque edge: a series of bright and dark 'fringes' appears, instead of the predicted full illumination or shadow.

For historical reasons diffraction phenomena are classified into two types: Fraunhofer and Fresnel diffraction. Fraunhofer diffraction treats cases where the source of light and the screen on which the pattern is observed are effectively at infinite distances from the intervening aperture. Thus, beams of light are parallel, so the wave front is plane, and the mathematical treatment of this type of diffraction is reasonably simple. Fresnel diffraction treats cases in which the source or the screen are at finite distances and therefore the light is divergent. This type of diffraction is easier to observe, but its mathematical explanation is considerably more complex.

However, a simple mathematical treatment of Fraunhofer

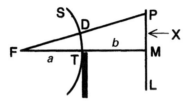

Fig. 12.4
Diffraction at a knife edge.

diffraction will serve to indicate some of the characteristics of diffraction effects. Diffraction at a knife edge is of most practical significance in radio propagation. The edge of any solid object approximates to it, and it is even quite a good approximation to the case of a radio wave passing over a ridge in the terrain.

Consider a flow of radio quanta towards the knife edge; for simplicity of analysis we assume a cylindrical front S (Fig. 12.4). How does the density of photons vary in the plane PML? The point P is x above M and we calculate the flow of quanta to this point.

The front can be considered as a series of horizontal strips, known as **Fresnel zones**, which are parallel to the knife edge, each strip having a lower boundary line half a wavelength further from P (Fig. 12.5). If the sth zone is distant r_s from P at its lower edge, the

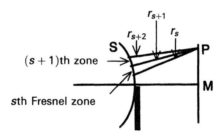

Fig. 12.5
The Fresnel zones.

$(s + 1)$th is distant $r_{(s+1)}$ and so on, then

$$(r_{s+1} = r_s + \lambda/2)_{\text{all } s}$$

Thus, every Fresnel zone is of approximately the same area, so that the same number of photons is passing through it, and each is half a wavelength further than its neighbour from P. If we consider the wave functions of the individual quanta in each Fresnel zone, clearly each individual one will have a neighbour in the next zone which is further from P by a half wavelength, and thus the wave functions of each such pair will arrive at P in antiphase, and so will cancel. (The difference in distance from P between adjacent pairs of zones is very small, therefore to good approximation any difference due to different spreading out of the quanta may be neglected.) It follows that if the number of Fresnel zones in the wave front is even, the wave intensity at P will be zero, but if odd it will be a maximum.

The series of zones on the side DS (Fig. 12.4) is unobstructed, so it will give half the amplitude of the whole wave front. On the side DT the wave is incomplete, being partly blocked off by the knife edge. As above, if there is an odd number of zones the result at P is a maximum, and if even then it is a minimum.

Assuming x small compared with b

$$\text{PT} = \sqrt{b^2 + x^2} \approx b + \frac{x^2}{2b}$$

similarly

$$\text{PF} \approx a + b + \frac{x^2}{2(a+b)}$$

so

$$\text{PD} = \text{PF} - a \approx b + \frac{x^2}{2(a+b)}$$

For a maximum at P, PT − PD = $n\lambda$ (resulting in an odd number of zones) which gives

$$\frac{x^2}{2b} - \frac{x^2}{2(a+b)} = n\lambda$$

or

$$\frac{ax^2}{2b(a+b)} = n\lambda$$

so

$$x_{max} = \sqrt{\frac{2n\lambda b}{a}(a+b)} \qquad (12.5)$$

Similarly for a minimum

$$x_{min} = \sqrt{\frac{(2n+1)\lambda b}{a}(a+b)} \qquad (12.6)$$

Thus, if, for example, b = 10 m, a = 10 km and λ = 1 m, maxima occur at x = $4.47\sqrt{n}$ m, that is 4.47, 6.32, 7.5 m, and so on, with minima between them (Fig. 12.6).

If P is below M all the zones are obscured on one side of D, and the

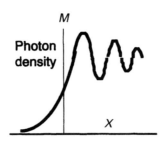

Fig. 12.6
Variation of photon density (= power density) at a knife edge.

first on the other side starts some distance from D. Thus the intensity diminishes steadily from M to zero over a little distance.

Note that the shorter the wavelength the smaller the dimensions of the diffraction patterns. Thus for maxima, from eqn (12.5) above, if X, A, B are x, a, b measured in wavelengths (that is, $X = x/\lambda$ and so on) then for maxima

$$X_{\max} = \sqrt{\frac{2nB}{A}(A + B)} \qquad (12.7)$$

and similarly for minima. Thus, the wavelength is a scaling term throughout; if the wavelength is reduced by a certain factor all the diffraction effects will physically scale down in size by the same factor. It is the position relative to the edge of objects measured in wavelengths which determines the diffraction effects, so the size of diffraction 'fringes' around objects depends on the wavelength.

The practical consequences of diffraction for radio propagation depend on the size of the obstacles relative to the wavelength of the quanta concerned. There is diffraction even around major terrain features such as hills or mountains by ELF, VLF and LF waves, with wavelengths measured in many kilometres, but shadowing is increasingly severe for radio transmissions with shorter waves than a few hundred metres length, and hence in the MF, HF and higher frequency bands.

The characteristic dimension of human beings is 1.5–2 m and few of their open-air artefacts are more than an order of magnitude different from this, that is most are less than 20 m in dimension but more than 0.15 m. In a flat-terrain built environment (suburbs, city) there will consequently be little shadowing of waves longer than about 10 m ($f < 30$ MHz), since diffraction around all objects will largely fill in shadows. Thus, in the HF bands and below (in frequency), shadowing by built structures is unlikely. At VHF there will be serious shadowing only by substantial buildings and large vehicles, but still with significant signal received some metres within the geometrical shadow area. At UHF, smaller objects (ordinary cars, road signs and other street furniture) will create shadows and it will be necessary to approach within the order of a metre of the geometrical shadow edge to receive signals. The effect grows even more marked as the wavelength gets shorter, until with millimetre waves (EHF) virtually all radio-opaque objects in the human size range throw deep and sharp-edged radio shadows.

12.4 Diffraction loss over terrain features

If a radio link is obstructed by a terrain feature (such as a ridge, treated as a 'knife edge') the signal does not fall at once to zero in the geometric shadow, nor does it immediately reach a uniform maximum value out of the shadow, as is evident from the preceding section. It is clear that if the wave path is above the knife edge there will still be possible signal variation due to diffraction, although this effect will diminish rapidly the higher the path. If the nominal signal path is below the knife edge the signal will be heavily attenuated but not zero, and the reduced number of quanta that do get through may still be used for communication. Calculation of diffraction loss over terrain features is therefore of considerable practical importance although analytically fairly intractable. A more detailed analysis yields the following results.

If we wish to communicate between the base and station 1 (Fig. 12.7), to good approximation the path loss (in dB) = path loss without obstruction + additional loss (L). To estimate L is difficult,

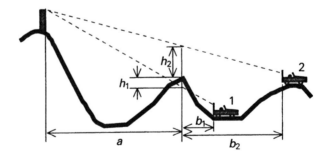

Fig. 12.7
Calculation of signal loss by diffraction over a ridge.

but the following empirical approach is widely used (Lee, 1989). Suppose that a, b are as shown; h is the obscuration clearance. Note that

1. The height of the obscuring ridge is corrected for refraction as explained in Section 12.1 above.

2. h is positive if the path is above the top of the obscuration and negative if below, i.e. negative for h_1 and positive for h_2.

Then a variable v is defined from eqn (12.5) as

$$v = \frac{h}{\lambda b} \cdot x_{max} = \frac{h}{\lambda b} \sqrt{\frac{2\lambda b}{a}(a+b)} \qquad (12.8)$$

So

v	Loss (voltage ratio) R_v
> 1	1
$0 < v < 1$	$0.5 + 6v$
$-2.4 < v < 0$	$0.4 - \sqrt{0.12 - (0.4 + 0.1v)^2}$
$v < -2.4$	$-\dfrac{0.23}{v}$

from which $L = 20 \log R_v$ gives the decibel additional loss.

12.5 Radio propagation in a complex built environment

As already indicated, in all bands up to HF there is good penetration of radio waves into a built-up environment because the waves are large enough to diffract round most of the objects encountered. At VHF and above, although diffraction plays a significant role in radio propagation in the typically cluttered urban or suburban radio environment, reducing the obscuration by objects not too large compared with the wavelength, an even more important part is played by reflection. When there are many reflecting bodies in close proximity to the receiving antenna and no line of sight (or near line of sight) path exists, radio quanta are propagated entirely by reflection from buildings, and to a lesser degree vehicles, assisted by diffraction round them. This **scattering propagation** scenario is of great importance. In the near-to-ground environment on Earth it is the dominant mode of propagation at VHF, UHF and above.

A full analysis of scattering propagation is inevitably complex. The papers by Egli (1945), who was a pioneer of the subject, and also by Allsebrook and Parsons (1977), are deservedly regarded as classics, and still repay study. The theory is extensive and only the principal results will be summarized here. First we consider the case of transmissions as unmodulated carriers, so that only amplitude and phase variations are significant, and subsequently the case of pulse modulation will be reviewed, to bring out the time-domain significance of scattering propagation.

As might be expected, radio propagation which relies on the chance that photons will be reflected by objects which happen to be in the right location to bounce them to the receiving antenna cannot easily be described by solutions of Maxwell's equations. Although calculations can be performed for any particular configuration of transmitter, scattering sites and receiver, each is different from all others. However, when there is (a) no line of sight between the transmitter and receiver, and (b) many significant scatterers (at least five or six within 100 wavelengths), the propagation converges to a stable statistical model, sometimes referred to as Rayleigh fading, with characteristic both local and longer range signal variation.

> John William Strutt, third Baron Rayleigh (1842–1919), made contributions in every area of classical physics. At the outset of his career he made his name by explaining the blue colour of the sky. Later he produced a definitive study of sound and laid the foundations for the theory of black body radiation. A major support to him in all his work was Mrs Sidgewick, his long-time collaborator, but when Lord Rayleigh was awarded the Nobel Prize for physics in 1904 she received no recognition.

Are the conditions required for Rayleigh statistics likely to apply in the case of scattering radio propagation? The first (no line of sight path for quanta) is highly likely in most built-up environments, at least for receiving antennas at street level or not much above. What about the second? Experimental studies in a city environment (without line of sight propagation) show that the number of scatterers is Gaussianly distributed with a mean number of 32 and a standard deviation of 12 (US results, but unlikely to be much different in other parts of the developed world). Evidently the conditions necessary for Rayleigh signal statistics will be met very commonly.

So, given these conditions, what does the scattered radio signal look like? Over variations in position of up to a few hundred wavelengths the signal is characterized by a Rayleigh distribution of amplitude and random phase. Longer range signal variation may be characterized as a slow variation of the mean of the statistical distribution which represents the short-range signal variation. This is the key to the satisfactory description of scattered radio transmissions.

First the Rayleigh distribution, for local variations. Where all propagation is by scattering the probability density function $p(s)$ of the amplitude s of received signal is (Fig. 12.8)

$$p(s) = \frac{s}{\sigma^2} \exp\left(\frac{-s^2}{2\sigma^2}\right) \qquad (12.9)$$

where σ is the modal value of s (V). This is a very highly variable distribution (Fig. 12.9). For comparison, after being filtered into a narrow bandwidth, Gaussian noise also has a Rayleigh amplitude distribution. Thus, it is not unfair to say that the amplitude distribution of the received scattered signal is about as random as it could possibly be.

The probability that the received signal s is greater than some value S is (by simple integration over the limits 0 to S)

$$P(s \geq S) = \exp\left(\frac{-S^2}{2\sigma^2}\right) \qquad (12.10)$$

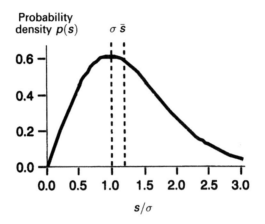

Fig. 12.8
The Rayleigh distribution.

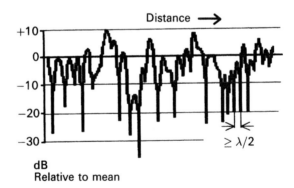

Fig. 12.9
How a Rayleigh distributed signal varies in amplitude with position of the receiving antenna.

Putting $P = 0.5$ in the above expression, it is easy to obtain the mean value of s, which is 1.1774σ ($+1.5$ dB relative to σ, bearing in mind that this is a voltage ratio). Also this cumulative distribution permits the calculation of the probability that the signal will exceed some minimum value (S) needed to give the required received signal-to-noise ratio. For 99% probability that the signal exceeds a

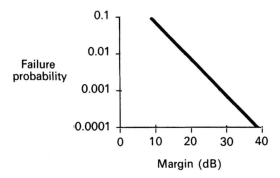

Fig. 12.10
Probability of failure to communicate as a function of 'safety margin'.

value S the value of σ is given by

$$\left(\frac{S}{\sigma}\right)^2 = -2\log_e(0.99)$$

and hence σ must be 17 dB greater than S. This is the reason why designers often set a mean signal 20 ($\approx 17 + 1.5$) dB above the minimum acceptable level as a 'safety margin'; it means that the signal will fall below the desired minimum in less than 1% of locations.

More generally, the probability that the signal will be of inadequate magnitude to give satisfactory communication is just one minus the probability calculated from eqn (12.10). In Fig. 12.10 this failure probability is plotted against the number of decibels by which the mean signal exceeds the threshold of the receiver. This graph can be interpreted in two complementary ways. A mobile-phone user will experience irregular breaks in reception due to the signal falling below the receiver threshold in the troughs of Rayleigh fades. Figure 12.10 therefore gives the proportion of the time that there will be no reception for a user who continues in motion. In a digital system it represents the minimum 'floor' below which the raw bit-error rate cannot fall.

For a stationary user the same figure is simply the probability that it will not prove possible to establish a radio link from the location currently occupied. So, for example, if a mobile phone user attempts to make a call, Fig. 12.10 gives the probability of being unable to do so. If the probability is 0.01, then one call in every 100 attempted from randomly chosen locations will be frustrated. As previously calculated, this occurs at a margin of 18.5 dB. The probability gets an order of magnitude less for every 10 dB further increase in mean signal.

Moving on to consider phase, in the Rayleigh distributed received signal all phases are equiprobable, so if ϕ is the phase, its probability density function is

$$p(\phi) = \frac{1}{2\pi} \qquad (12.11)$$

where ϕ is in radians.

Sharp phase changes occur in fading troughs, often resulting in spurious outputs in phase or frequency modulated systems.

12.6 Longer range signal variation

Within a few wavelengths of a fixed point the mean value of the received signal is approximately constant, so the Rayleigh distribution is a good approximation to the distribution of signal amplitude. Over longer ranges the variation of signal may be represented as a variation of mean value and hence σ. Obviously the mean signal strength will tend to fall as the distance from the transmitter increases. As we have seen, in free-space propagation an inverse square law applies but under conditions of scattering propagation the mean signal falls more rapidly with range, as a consequence of multipath effects. Indeed, this is the case under multipath conditions much less extreme than a true Rayleigh fading environment.

Recall the simple case of a single ground reflection (Fig. 11.7, earlier). If the distance between transmitter and receiver is $2d$ and the height of both above ground is h, the path length difference between a direct and reflected wave was shown to be

$$\Delta = 2d\left(1 + \frac{h^2}{2d^2}\right) - 2d = \frac{h^2}{d}$$

which produces a phase shift between the two wave functions arriving at the receiving antenna equal to

$$\phi = \pi + \frac{2\pi}{\lambda} \cdot \frac{h^2}{d}$$

Hence, assuming that the two received signals are of equal amplitude e

$$E_{\text{sum}} = e \cdot \cos(\omega t) + e \cdot \cos(\omega t + \phi)$$
$$= 2e \cdot \sin\left(\omega t + \frac{\pi h^2}{d\lambda}\right) \cdot \sin\left(\frac{\pi h^2}{d\lambda}\right)$$

This is simply the sum of the received sinusoids (with a phase shift) multiplied by the sine of the phase difference, and thus at the receiving antenna the received power will similarly be multiplied by the square of the same sine term, so using eqn (10.5) and bearing in mind that the range is $2d$

$$P_R = \frac{A_R G_T}{16\pi d^2} P_T \cdot \sin^2\left(\frac{\pi h^2}{d\lambda}\right)$$

In the near-to-ground case h is small, so the sine may be replaced by the angle giving

$$P_R = \frac{A_R G_T}{16\pi d^2} P_T \left(\frac{\pi h^2}{d\lambda}\right)^2$$
$$= \frac{\pi A_R G_T h^4}{16\lambda^2 d^4} P_T \propto \frac{1}{d^4}$$

So we may write

$$P_R = \frac{KA_R G_T}{d^4} P_T \qquad (12.12)$$

where K is a constant.

This is the **inverse fourth power law** which relates both the Rayleigh distribution mean and also its modal value σ to range from the transmitter. It is in sharp contrast to the inverse square law which applies in space, and demonstrates that the range of a transmitter which is obliged to depend on scattering propagation will be much reduced.

How well is it borne out in practice? If it were exact, the mean signal should fall by 40 dB for a 10:1 increase in range. This gradient has been measured, both for European and US cities, and results are mostly in the range 40 ± 3 dB, generally coming out on the high side in city centres and lower in suburban areas. However, there are some exceptions. At sidewalk level in downtown Manhattan, surrounded by closely spaced and exceptionally high buildings, a figure of 48 dB has been recorded. By contrast central Tokyo shows the unusually low figure of 31 dB, as a result of the topography of the city, surrounded by hills on which the radio transmitters are sited, so that line of sight transmissions are not infrequent. (Where propagation is partly by line of sight and partly by scattering it is called Ricean.) Even so, substantial deviation from the 40 dB figure is exceptional, and the inverse fourth power law remains a useful indicator of likely system performance in the absence of measured field statistics. Experimental results do back up the inverse fourth power law quite well.

By using the inverse fourth power law to model the variation in the value of σ, in a Rayleigh distribution it is a straightforward matter to predict the performance of terrestrial radio systems, particularly area coverage and co-channel interference (Gosling, 1978).

It is also important to consider the effects of scattering propagation in the time domain. This has been studied experimentally, particularly in the VHF and UHF bands, by using short pulse transmis-

At ground level **197**

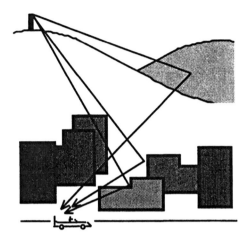

Fig. 12.11
The reflection from the hill would be strongest and arrive last.

sions and investigating their relative amplitudes and time of arrival at the receiver. As would be expected, the amplitude depends primarily on the reflecting area, while time of arrival is proportional to total path length, three microseconds for each kilometre. The first pulse to arrive may therefore not be the largest, since it may come from a relatively small reflector (Fig. 12.11).

Many studies have been reported of the spread of arrival amplitudes and times, which are obviously critically dependent on the density of scatterers in the physical environment of the receiving antenna. In the UHF band, for an open area delay spreads of 0.1 to 0.2 µs have been reported, in suburban areas the spread is likely to be 0.5 to 1 µs, whilst in urban areas it may exceed 3 µs. The figures quoted are not very frequency dependent until the millimetre wave bands are approached, when they tend to fall (nearer scatterers become more significant). Similarly larger values (more distant reflections) are recorded at VHF.

Delay spread becomes significant if the radio signals have digital modulation. If the delay spread is of the order of the inter-symbol interval, severe increases in raw error rates occur (i.e. at about

300 kbaud when the delay is 3 μs) because '0's and '1's received by different paths overlap. This effect may be usefully offset by the use of equalizers.

12.7 Trees

Some attention has been given to radio propagation through woodlands. This is particularly important to the military and also to rural cellular radio telephone users, although the presence of trees on the summit of a ridge which is acting as a diffraction knife edge in a transmission path cannot be overlooked.

This is a very complicated topic, with some odd effects. Wet trees absorb more strongly than dry ones. Also, there is a big difference between summer and winter foliage absorption with deciduous trees, but not, of course, with evergreens. Worse, for cellular radio at 900 MHz, half a wavelength corresponds to 16 cm (just over 6 in). If trees have leaves with this characteristic dimension they will absorb strongly in the wet, due to the effect of resonance. There are many such trees, notably certain conifers which have needles of this length.

Other significant factors naturally include the closeness with which trees are planted. If trees are well spaced, for example on the African Low Veldt, they can themselves act as scatterers and enhance the available signal between them, but if they are densely packed, as in a tropical rain forest, much of the propagation path is through the tree cover and the absorption is severe. It has been shown that where radio transmissions propagate through continuous heavy tree cover an inverse sixth power law replaces the normal near-to-ground inverse fourth power law, and this is often used to estimate path losses.

12.8 The effects of buildings

So much for radio services to the out-of-doors user. However, to give true universality of communications it is now increasingly required to be able to use portable radio equipment inside buildings. This involves two possible modes of operation: the reception of transmissions originating within the building and communication with stations outside, which involves penetration by the radio transmissions through the walls.

Building penetration by radio quanta has been most thoroughly studied at 800–900 MHz, for cellular radio telephone services, with less intensive investigations at other frequencies. Needless to say, the means of building construction is critically important. An all-metal building with no windows hardly admits any radio energy at all, but fortunately except for military installations and semiconductor foundries very few buildings are constructed this way. Even so the techniques of building are dominant.

For example, at 800 MHz the path loss difference inside and outside a building at first floor level in Tokyo has been measured as 26 dB, but similar studies in Chicago yield 15 dB. This is because typically much more metal (such as mesh) is used for building in earthquake-prone Japan than in Chicago, where earthquake risks are slight. In Los Angeles, where the earthquake risk is intermediate, the measured loss is 20 dB. Results of measurements are only meaningful, therefore, when related to building techniques.

Experimental studies in the UK (Turkmani *et al.*, 1987) show a penetration loss in the range from a few decibels up to 20 dB on the ground floor, tending to decrease with frequency (presumably due to better penetration through windows and other apertures), loss decreasing on higher floors, typically by 2 dB per floor, and Rayleigh distributed signals within the building. These conclusions have been confirmed for European building styles by several studies, and may be taken as valid for the VHF and UHF bands. In the SHF, particularly above 10 GHz, the transparency of buildings to radio signals decreases, and for millimetre waves (EHF) there is virtually no penetration.

Because of the interest in, for example, radio-based cordless PABXs (internal telephone systems), telephones and other cordless systems, the propagation of radio waves within buildings has also been widely investigated in recent years. Studies have been mostly conducted at UHF and above, because of the need for small antennas inside buildings, better penetration of apertures within buildings at shorter wavelengths and less restrictive frequency assignment availability from the radio regulatory authorities.

As might be expected, room-to-room propagation is by scattering, with the geometry such that in general all signals will have undergone several reflections before being received. This makes for a rapid rate of signal attenuation with distance, along with Rayleigh statistics of signal variation. Wavelengths should ideally be small compared with door and corridor dimensions, so frequencies below 300 MHz result in less uniform coverage. With low powers there is little energy radiated outside the building from transmitters within, which is good for spectrum conservation and yields a degree of privacy.

An indoor channel is found to have the advantage of relatively small delay spread (because of the close proximity of reflecting surfaces), limiting the undesirable effects of multipath transmission. Thus, acceptable performance (raw BER $< 10^{-4}$) can usually be obtained without the need for channel equalization at bit rates up to 30 Mb/s, which allows the design of low-cost user terminals for many applications.

These conclusions remain valid up to about 20 GHz. Above this frequency, internal partitions become opaque and radio shadows are not significantly filled in by diffraction. Use of millimetre waves within buildings concentrates on covering each room by its own transmitter, located on the ceiling and transmitting down on to the user population, to put radio shadows in unimportant locations. If room-to-room communication is required, cable-linked active repeaters are used. This approach achieves excellent privacy and spectrum conservation, since transmissions do not escape the rooms. However, there is challenging competition from infrared systems, which provide a similar service.

Problems

1. What do you understand by multipath propagation and what are its main consequences for radio transmission? Show how in the case of VHF point–point transmissions this effect can cause received signal strength to be unusually sensitive to antenna height. If a transmitter at a height of 100 m above a smooth reflecting Earth communicates with a receiving antenna at a similar height and 10 km distant, what will be the coherence bandwidth? [150 MHz]

2. In a scattering radio environment a mobile receiver moves around in a small area whilst receiving signals from a wanted transmitter A and an interfering transmitter B. If B is twice as distant as A what is the probability that the signal from A will exceed that from B by at least 10 dB? [0.88]

3. A police mobile receiver requires -120 dBm to function correctly. In the vicinity of a certain cross-roads 20 km from its base it does so 97% of the time. The police vehicle gives chase and moves out to a housing estate 30 km from its base. As it moves over a short distance there, what proportion of the time will it receive satisfactorily? [46%] Assume a scattering propagation environment.

4. Within a building of apparently normal construction there is high attenuation of radio signals in the UHF band originating outside. What would you suspect as the cause?

5. An army patrol carries VHF radio equipment which gives satisfactory communication to a distance of 20 km from their base in the scattering environment. When in open country adjacent to an extensive heavily wooded area they come under enemy fire and decide to take cover. They enter the woods at a distance 10 km from their base and are obliged to move radially away. It is raining. How far into the wood will they lose radio communication? [about 6 km] What could they do to restore it?

CHAPTER 13

THE LONG HAUL

Although the most commercially important applications of radio technology at present are for short- and medium-range applications (from cellular phones to terrestrial broadcasting), long-haul communication systems remain important. It is true that intercontinental traffic is increasingly handled by optical fibre, but there are many situations in which radio remains the carrier of choice. An obvious example is for maritime communications, and indeed all mobile communications requirements must be met, for at least part of the transmission path, by using radio. But, in addition, there can be compelling reasons for using radio even where the stations are at a fixed location. Obviously this is true in more remote places where there is no access to the optical fibre network, also where temporary communications must be set up and broken down quickly (as on civil engineering projects) and not least where the physical security of the link must be high enough to survive even disaster situations. These considerations dictate radio solutions in many civil situations, and often dominate emergency service and military thinking.

For radio links up to 100 km or so, the technology already described is adequate but, even so, proximity to the Earth's surface does lead to severe losses over longer distances, which is a problem for the long-haul communicator. Nevertheless, the earliest long-haul systems used surface propagation.

13.1 Surface waves

On any large conducting surface, such as the Earth, it becomes possible to propagate radio quanta characterized by **surface waves** (Fig. 13.1). (Surface waves are sometimes called ground waves, but this term is used in another sense also, and is therefore undesirable.) The fields associated with the passing quanta induce currents in the conducting plane (the Earth's surface) which play a critical part in the propagation mechanism. For VLF to MF frequencies this is the only practicable transmission mode, since to be even approximately clear of Earth effects it would be necessary to elevate antennas by several wavelengths, which is not practicable. Instead, vertical Marconi antennas (see Section 6.5, earlier) are invariably adopted for transmission, and, except at the higher MF frequencies, are always much shorter than a quarter wavelength, leading to poor efficiency. They must be tuned to resonance by an ATU. At the receiver, either similar Marconi antennas are used or ferrite rod antennas (see Section 8.1 earlier).

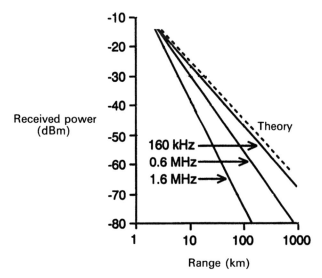

Fig. 13.1
Surface wave propagation.

Surface waves are launched by vertically polarized antennas over the Earth's surface. The Earth is spherical, so propagation over this curved surface depends upon diffraction. The mathematics of this mode of propagation is complicated, and has been treated by Sommerfeld and others. It leads to a model, with a perfectly conducting Earth, in which received power falls inversely with the square of distance. In practice the Earth is not perfectly conducting at these frequencies, so the rate of fall of signal is greater. At VLF, surface waves are launched but propagation is more complex, due to trapping of the waves between the Earth's surface and the ionosphere (see below). This leads to a type of wave-guide propagation. Surface waves are characterized by excellent phase and amplitude stability at the receiver, due to the simple propagation mechanism.

Using very high power for hand-speed Morse code transmissions, from 1910–27 or so, this was the only intercontinental radio communications technology. In 1907, Marconi's collaborators had reported to a London IEE meeting that reliable transatlantic communication was practicable in the VLF band with a transmitter power of 600 horsepower (450 kW), yielding a single telegraph channel capable of transmitting at up to 25 words per minute (about 20 bits/s). Extreme power transmitting stations along these lines were actually built, notably at Clifden (Ireland) (subsequently destroyed by the IRA) but with the introduction of HF technology in the mid-1920s a much cheaper alternative for intercontinental communications became available, allowing telephony as well as telegraphy.

Surface wave propagation continues to be used today, but only where its unique characteristics of extreme stability and reliability are essential. In view of the limited bandwidth available (the whole band is only 27 kHz wide, and a mid-band station with an antenna Q-factor as low as 10 has a bandwidth only a little over a kHz) the VLF band is limited to slow data rate transmissions, but they can be received worldwide. Used for standard frequency and time transmissions (e.g. GBR on 16 kHz and GBZ on 19.6 kHz, both at Rugby), and for the Omega worldwide positioning and navigation system (at around 11 kHz), it also has military interest for communicating with ships and submarines. The US flying command centre, a converted jet airliner from which the US president would control a nuclear war, relies on VLF transmission from a trailing wire antenna as its 'last ditch' communications medium.

Surface waves are also the basis of all MF and LF broadcasting (up to 300 km range LF, 100 km MF), as well as LF radio positioning systems. LF was the original AM broadcasting band in Europe (but not in the USA, which opted for MF) and is still used primarily for that purpose. Because of the higher centre frequencies the bandwidths of equipments are proportionately higher than at VLF, which enabled Reginald Fessenden to pioneer their use for speech and music broadcasting.

> Reginald Aubrey Fessenden (1866–1932), a Canadian, began radio research at the University of Pittsburgh. In 1906 he broadcast the first programme of speech and music ever transmitted, using a high-frequency alternator as his radio power source. Fessenden also invented the heterodyne receiver, from which Edwin Armstrong (1890–1954) evolved the modern superheterodyne (1918).

The surface wave propagation mode is used to a very limited degree at HF for short-range (a few kilometres) military and rural radiotelephone services, using vertical whip antennas. At VHF and above, the attainable range is too small to be of any use.

13.2 Sky waves

Although intercontinental radio communication began using surface wave propagation, this proved not to be an ideal approach for many potential applications. Losses over the Earth's surface are increasingly severe as frequency is increased, limiting intercontinental communication to the VLF band, where very little bandwidth is available and resonant circuits (and antennas) in the transmitter restrict data rates to very low values even when the Q-factor is low. The only way to escape from these limitations is to direct the radio energy away from the Earth's surface into space, where losses are very much lower provided that we stay below frequencies where atmospheric absorption is significant. This is **sky wave** communication; its essence consists of directing the main lobe of the transmitting antenna well above the horizon, so that the quanta are launched on a path away from the Earth's surface, and then using some means to return the radio energy back to the ground close to the intended receiving station.

We have already seen that ducts, caused by inversion layers, are naturally occurring atmospheric phenomena which will do just this for transmissions in the VHF and lower UHF bands (see Section 11.2, earlier). However, these are chance meteorological phenomena which do not give reliable means of communication (other than in a very few exceptional locations). Much more certain methods of securing the return of the sky wave radio transmissions to Earth exist and form the basis of several radio communications technologies currently in widespread use.

What of the atmosphere itself? Can collisions between photons and air molecules deflect some of the radio energy back to the ground? Although the troposphere is normally regarded as transparent to electromagnetic radiation up to around 20 GHz, at frequencies between 1 and 5 GHz in the troposphere there is sufficient Thomson scattering of radio quanta to make propagation by this means possible. This **tropospheric scatter** consequently provides a means of long-distance (100–1000 km) communication, but with considerable path loss (>140 dB). This is because the cross-section of air molecules for collision by photons is very low. The technology, therefore, depends on transmitting very high powers from high-gain antennas, launching transmissions at low angle ($< 5°$) to equally high-gain receiving antennas. Although 'tropo scatter' can give reliable links, it has now been eclipsed by other communication technologies. The design of such systems is largely empirical.

Current sky wave communication systems use one of three techniques for returning the signals to earth: **meteor scatter**, **ionospheric propagation** and **artificial satellites**.

13.3 Meteor scatter

A **meteor**, often called a shooting or falling star, is seen as the streak of light across the night sky produced by the vaporization of interplanetary particles as they enter the atmosphere. A few larger meteors are not completely vaporized, and the remnants that reach the Earth's surface are called **meteorites**, and are believed to add as much as 1000 tonnes to the Earth's mass every day. Although a few meteors can be seen on any clear night, especially after midnight, during certain times so many are visible that they are termed meteor showers. Records of such showers go back to the eleventh century.

Meteors are thought to be the debris from comets, which leave a trail of matter behind them as they orbit the Sun. As the Earth passes through this cloud of debris, many particles, on average about the size of a grain of sand, enter the atmosphere at speeds up to 100 km/s. Air friction causes them to vaporize, creating a large elongated trail of very hot gas, the visible shooting star, in which the temperature is high enough for the outer electrons to be stripped off atoms in the gas, leaving clouds of positive ions and free electrons. The process of heating begins some 80 km out and the burning-up is almost always complete by the tropopause, at 20 km. The very large cloud of **ionized gas** produced is able to refract and reflect radio quanta, mostly because the free electrons have a high cross-section for collision with radio quanta in the VHF band. However, the positive ions and free electrons readily recombine, so the phenomenon does not last; as soon as it is created the ionization begins to die away. This **recombination** happens faster the higher the atmospheric pressure, because high pressure brings ions and electrons closer together on average. So trails formed at higher altitudes will last longer than lower ones, although this is somewhat balanced by the trails at lower altitudes (where heating is fiercer) being more intense initially. Meteor scatter radio systems work by directing a radio sky wave at a meteor trail, which reflects it back to Earth (Fig. 13.2).

Because the exact location of the incoming meteor is not known, either in position or altitude, by covering a relatively large area of sky using a relatively wide antenna lobe (at both transmitter and

208 Radio Antennas and Propagation

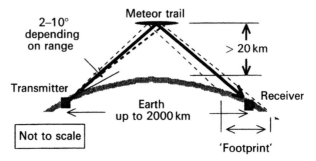

Fig. 13.2
Meteor scatter propagation.

receiver) it is possible to maximize the chance of intercepting an ionized trail. At the same time a large aperture is also desirable to ease the power requirement at the transmitter and it must be easily possible to alter the angle of elevation of the antenna, so as to communicate at different ranges. Putting these considerations together, the optimal antenna choice is a VHF Yagi array, which has large aperture whilst retaining a wide main lobe.

Because they are launched towards the reflector over a range of angles, the returning radio quanta are spread out over a '**footprint**' on the Earth's surface around the designated target point. Due to the geometry, at ground level there is nothing to be intercepted in the **skip area** between the transmitter and the 'footprint' where the signal returns to Earth. In consequence, meteor scatter has good privacy (attractive to the military) and gives rise to minimal spectrum pollution because it cannot cause interference outside the 'footprint'. Systems operate over paths of up to 2000 km in length, beyond which the geometry of the system dictates so low an angle of elevation for the transmission that losses due to the surface and terrain features become excessive, although hilltop sites for transmitter and receiver can help (Fig. 13.3). If the link is to be used exclusively for long-range transmission, the antenna main lobe width can be very much smaller, with consequent improvement in gain and usable trail duration. Systems designed to cover many differing transmission paths must have a wider lobe. There is little interest in meteor systems for ranges of 100 km and less, to which

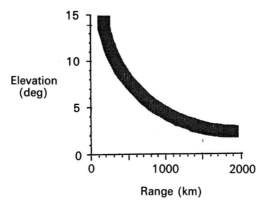

Fig. 13.3
Antenna elevation versus range – estimated due to uncertain trail position.

normal point-point radio techniques can be applied, although there may be classified military applications.

The principal difficulty of meteor scatter communication is that the transmission path, although utterly reliable in the long run, is only present intermittently. Over a given path the probability density function for a wait t before a path is open is $p(t)$, where approximately

$$p(t) = \frac{1}{\tau} \exp\left(\frac{-t}{\tau}\right)$$

The probability of a channel appearing in a time T is thus

$$P(T) = \int_0^T p(t) \cdot dt = \left[-\exp\left(\frac{-t}{\tau}\right)\right]_0^T = 1 - \exp\left(\frac{-T}{\tau}\right) \quad (13.1)$$

In middle latitudes the value of the average wait is typically 2.5 min (Fig. 13.4).

Once a path is open there is a similar exponential distribution of useful channel 'life', that is the time until the signal-to-noise ratio

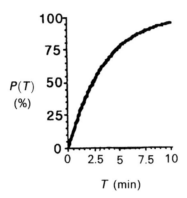

Fig. 13.4
Probability that a channel will become available versus wait.

has fallen too low to sustain the required data transmission rate. Obviously this is longer the higher the transmitter power and the lower the minimum acceptable signal level. However, because the ionization decays exponentially, raising transmitter power only gives a modest extension of the 'window' duration, and the typical transmitter power (100–500 W) is mostly determined by what is available at moderate cost. Higher altitude trails recombine more slowly, and so last longer, but they are less frequent. With all the variable factors, a typical mean for the usable channel duration is only 0.5 s.

Even so, whilst a channel is active, data can be passed at a high rate (up to at least 100 kb/s, depending on the transmitter and receiver hardware details) but the short duration of the functioning life of each meteor trail means that the average message capacity over a 24 h period is commonly less than 200 bits/s in each direction (allowing for system and control transmission, as explained below). However, even this would amount to 1.6 Mb/day in each direction, which is enough for many e-mail, data monitoring and short messaging requirements.

Meteor scatter systems are an example of burst transmission 'store and forward' operation, in which digital traffic is stored until the

Table 13.1 Meteor scatter operating protocol

1. One station is designated as master, the other as slave
2. The master station transmits probe signals at regular intervals, usually about ten per second, and listens in between for responses
3. The slave station listens until it receives a probe signal at acceptable signal-to-noise ratio, then quickly responds with a 'handshake' transmission, during which it indicates whether it has traffic to pass
4. The master then transmits, either passing its own traffic or calling for the slave's traffic
5. Whichever station is originating traffic does so in short blocks, with return acknowledgement from the other station in between. When the master receives no acknowledgement (or no new data block after sending an acknowledgement) it reverts to (2) above

channel opens and then rapidly forwarded to its destination. Stations using meteor scatter are therefore invariably computer controlled (Table 13.1).

Meteor scatter only achieves a modest average transmission rate, but has a number of advantages which result in its use in niche applications to which these particularly apply. Perhaps surprisingly, the most important of its virtues is reliability: the channel will always be there, sooner or later, because new meteorites are always entering the atmosphere, many thousands every day. This is certain, not affected by ionospheric conditions, weather and so on, and there are only minor changes from time to time, as the Earth passes through meteor showers. No space vehicle launching capability is required for this technology, as with satellites, and it is consequently cheap. At each end of a meteor scatter link little more is required than the equipment typically found in an amateur radio station, including a desk-top computer. Used remote from telecommunication services, particularly in underdeveloped areas, meteor scatter is valued wherever a moderate data transmission rate with delays of up to a few minutes is acceptable (for example, in meteorological, water resource and environmental management systems).

13.4 Ionospheric propagation

Although its existence had long been suspected, notably by Nikola Tesla (1856–1943) (who unsuccessfully built a tower to try to make contact with it), Edward Appleton first obtained observational evidence for the **ionosphere** in 1925. The ionosphere is a consequence of the ionization of the upper atmosphere by energetic electromagnetic quanta emitted from the Sun, mostly ultraviolet but also soft X-rays. These quanta collide with air molecules or atoms and strip off their outer electrons, leaving them positively ionized in a sea of free electrons. However, at a rate depending on local atmospheric pressure, the process of recombination is going on all the time, so that the degree of ionization is a consequence of a dynamic equilibrium between continuing ionization and recombination. It is therefore greatest over a particular part of the Earth's surface in summer and daytime, least in winter and at night.

The ionosphere is subdivided into: the **D layer** between 60 and 85 km in altitude, which disappears at night due to the rapid

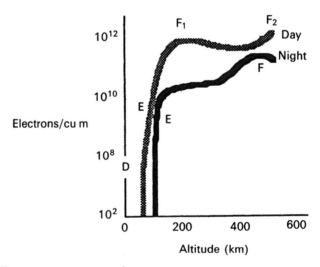

Fig. 13.5
Typical variation of free electron density with altitude by day and night.

recombination in the relatively high pressure, the **E layer** (sometimes called the Heaviside layer) which is at an altitude between 85 and 140 km, and the **F_1 and F_2 layers** which are found above 140 km and merge at night (they were formerly both called the Appleton layer) (Fig. 13.5). In the D and E layers it is air molecules which are ionized, but in the F layers pressure is so low that atmospheric gases exist primarily in atomic form, so that the positive ions are atoms. The ionized gases in these layers can refract and reflect radio quanta in just the same way as described above for meteor trails.

Ionospheric recorders at stations on the Earth's surface give information on the lower regions of the ionosphere. An exploring signal (usually in the HF band) transmitted vertically upward is reflected downward by the ionosphere to a nearby receiver. The height of the reflecting level is obtained from the total transit time of the signal. However, the upper half of the ionosphere is probably now better known than the lower, due to the placing of ionospheric recorders on satellites, beginning with the Canadian *Alouette*.

The ionosphere has two effects on radio quanta: absorption and refraction. By collisions with the radio quanta, free electrons receive energy which is subsequently dissipated in further collisions with electrons, positive ions or gas atoms. As a result energy is lost from a radio transmission passing through a region of high density of free electrons. The cross-section of an electron for collision with a photon, and hence the probability of such an event, depends on the energy of the photon. Alternatively, the free electron, which has been raised to a higher energy state by the colliding photon, before it has had time to give up its energy by further collisions, may relax back into its initial energy state releasing a new photon with energy identical to the original. The effect is exactly equivalent to a deflection of the photon. The path of the stream of radio quanta will consequently be redirected, and the result can be that the radio transmission is turned back towards the Earth, as a consequence of this collision process (Fig. 13.6). Thus, the photons will sometimes escape, if their energy is high enough or the free electron density low enough, and at others they will be returned to Earth. What the outcome will be depends on the electron cross-section for collision, the length of the path in the ionized layer and the density of free

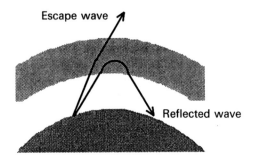

Fig. 13.6
The ionosphere may reflect the stream of radio quanta back to Earth, but if their energy is high enough they can escape.

electrons, factors which jointly determine how many collisions are likely to take place.

We begin the analysis with the case of a radio transmission directed vertically upward at an ionized layer. A stream of radio quanta passes through the ionized region; consider a thin slice of area A (Fig. 13.7). The probability that there will be a collision depends on the cross-section for collision of the free electrons. What is this

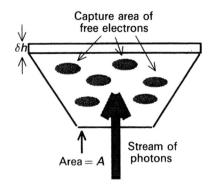

Fig. 13.7
A radio transmission comprising a stream of radio photons traverses a thin slice of an ionized layer.

cross-section for collision? This is the term preferred by particle physicists, but we are already familiar with the idea under another name. It is simply the area within which a passing radio quantum will be captured by ('collide with') a free electron. But we are already well aware of the possibility that electrons in receiving antennas can capture quanta; what we are considering here is just the capture area of an individual electron, which is therefore of the form $k\lambda^2$ where k is some constant. If the density of free electrons is N, the number in the slice is $NA \cdot \delta h$ so the probability of a collision in the slice is

$$p = \frac{\text{total electron capture area}}{A}$$

$$= \frac{k\lambda^2 NA \cdot \delta h}{A} = k\lambda^2 N \cdot \delta h$$

In the limiting case, the value of p must exceed a certain minimum in order that there shall be enough collisions to return the quanta to Earth, so we may write

$$p > p_{\min}$$

Hence

$$\lambda^2 > \frac{p_{\min}}{kN \cdot \delta h}$$

or, in frequency terms

$$f < f_c \text{ where } f_c \propto \sqrt{N} \tag{13.2}$$

Here f_c is the **critical frequency**, which is the frequency at which the return to Earth of the radio quanta just fails. As might be expected, extensive experimental studies of the critical frequency for the various ionospheric layers have been carried out in many geographical locations and over many years. In the ordinary way it increases steadily with the height of the layer, since the high layers have lower recombination, and hence higher electron densities, than the lower. The critical frequency shows marked diurnal variation (Fig. 13.8), rising during the day (when electron density increases due to the Sun's action) and falling at night (when it falls due to

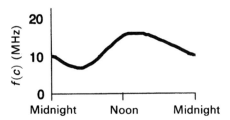

Fig. 13.8
Typical diurnal variation of critical frequency.

recombination). It is also dependent on the season, being highest in summer, and on the strength of the Sun's activity, which varies over an 11-year cycle.

When the signal is not launched vertically but at a lower angle (Fig. 13.9), the quanta pass through a longer, slant path in the ionized layer, increasing the chance of a collision with the path length. The escape frequency at this angle is consequently increased, leading to a **maximum usable frequency** (muf) for communication given by

$$f_m = \frac{f_c}{\sin \theta} \tag{13.3}$$

But the launch angle required is determined by the communication path length (Fig. 13.10) (the lower, the further). If $2d$ is the great-circle distance between two sites (that is, the curved path measured over the Earth's surface) then if the radius of the Earth is a, the angle subtended at the centre of the Earth by the path is $2d/a$. Thus the mid-point apparent rise in the Earth's surface is

$$s = a(1 - \cos d/a)$$

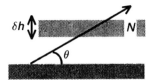

Fig. 13.9
A sky wave transmission at a launch angle θ.

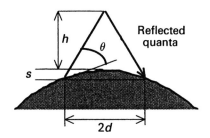

Fig. 13.10
Geometry of the sky wave.

The base angles of the triangle formed by the propagation path and the straight line between the sites are

$$\alpha = \tan^{-1}\left[\frac{h + a(1 - \cos d/a)}{a \sin d/a}\right]$$

and (neglecting any effect of lower atmosphere refraction) the launching elevation of the radio signal is

$$\theta = \alpha - d/a = \tan^{-1}\left[\frac{h/a + (1 - \cos d/a)}{\sin d/a}\right] - d/a$$

In practice h can only be approximated, since it varies with time of day and the seasons, and anyway the lobes of HF antennas used for ionospheric communication are wide enough to accommodate a few degrees of error in elevation angle, so it is acceptable to use the simplified form

$$\theta = \tan^{-1}(h/d) - d/2a \qquad (13.4)$$

However, there are limits on the usable angle of elevation from the ground. Less than 5° leads to large losses (as with surface waves) and the highest angle practicable with F layer reflection is about 74°.

This analysis assumes that the radio energy is projected into space by the transmitting antenna, reflected by one of the layers of the ionosphere, and then returns directly to the receiver. In between is the skip distance, in which the signal path is not near enough to Earth for it to be received at all. However, the situation may be more complicated than this because, as we have seen, both land and (particularly) sea are themselves effective reflectors for radio waves. In the HF band, where the wavelength is between 10 and 100 m, for the most part the reflecting surface is relatively smooth. Near-specular reflection from the ground is therefore commonplace, making possible **multiple hop transmission**. In the two-hop case, the transmission is reflected from the ionosphere back to ground, reflected there back up to the ionosphere again and thence down once more to the receiver. Three and even four hops are possible but rare.

Designing for these multiple hop paths is complex. Obviously, so far as elevation angle is concerned, in the two-hop case the graph of Fig. 13.11 applies, but with the range axis doubled. The transmission frequency must, of course, be safely below the muf at both ionosphere reflection locations, and they may not be the same, since one might be in day when the other is in night. Sometimes either one- or two-hop solutions to the propagation requirement can be found, using different angles of elevation of the antenna. It is

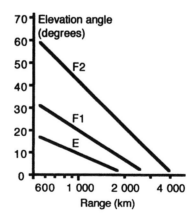

Fig. 13.11
Approximate antenna elevation for one-hop ionospheric propagation, plotted against range.

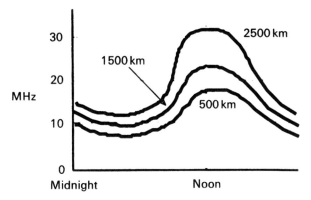

Fig. 13.12
Typical diurnal variation of muf over a variety of path lengths.

impossible to say which will be best without knowing the mufs for the reflection locations.

Given the angle of launch, and supposing the critical frequencies are also known (not locally to the transmitter but for the parts of the ionosphere where the reflection is expected to take place) the mufs can be calculated for both single and multiple hop cases. For a typical noon critical frequency of 15 MHz this will range from 16 MHz at an antenna elevation of 70°, up to 86 MHz for an elevation of 10°. All higher frequency transmissions will escape into space. At night these values may fall by half (Fig. 13.12). Evidently, ionospheric propagation is only possible up to the HF band, and sometimes the very lowest fringes of the VHF under specially favourable conditions.

Because the gain of a transmitting antenna is proportional to the square of frequency in the usual case of constant physical size, it is desirable to work as near as possible to the muf. In practice, the prediction of the muf is not sufficiently reliable to allow safe working much above 85% of its value. Intercontinental transmissions may be in day over some parts of the path and night over others; it is the lowest muf over the path which sets the limit on usable frequency. A **least usable frequency** (LUF) can also be defined, which depends on antenna and receiver parameters and transmitter power, and is the least frequency at which an acceptable signal-to-noise ratio is obtained at the receiver.

13.5 Ionospheric propagation in practice

The D layer, at a height below 80 km, is characterized by a very high rate of recombination due to the relatively high air pressure, so it exists only in daytime and vanishes quickly at dusk. For this reason the electron density is low, as therefore is its critical frequency, so it is penetrated by all radio transmissions above the MF band. When the D region is present, radio energy of long enough wavelength to interact with the electrons present is quickly dissipated via collisions by electrons with the many surrounding air molecules. This is what happens to MF signals. They are rapidly attenuated by the D layer and ionospheric reflection is virtually non-existent; in this band only the surface wave can be received strongly during daytime (Fig. 13.13). It is therefore around surface wave propagation that the service is designed.

At night (particularly in winter) the D layer becomes vestigial due to recombination, and the electron concentration required for MF reflection is now found about 50 km higher, in the E layer, where the atmosphere is so thin that much less absorption takes place. As a consequence many sky wave transmissions (ionospherically

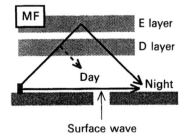

Fig. 13.13
By day the D layer absorbs the MF sky wave, which at night propagates via the E layer.

propagated via the E layer) come in from distant transmitters, filling the band with signals and greatly increasing the probability of interference between transmissions. The winter nightly cacophony on the MF (medium wave) broadcast band is a result of this unwanted sky wave propagation, and is evident to anybody who cares to turn on a broadcast receiver. To the broadcasting engineer, MF ionospheric propagation is wholly negative, spoiling night-time winter service. One reason for top-loading MF antennas is to reduce sky wave radiation. So far as LF and still lower frequencies are concerned, even at its weakest the D layer has sufficient electrons to trap them, and there is no sky wave transmission at all.

By contrast, at higher frequencies, where surface wave propagation is of much less importance (except to the military and for rural radiotelephone), ionospheric propagation is exploited positively. Intercontinental transmissions are almost always possible at some frequency in the HF band (3–30 MHz). Radio quanta in this energy range are reflected by the E or F regions and also by the Earth's surface, so that multiple hops are not uncommon, connecting points on opposite sides of the globe. Still shorter waves (higher VHF, UHF and above) will be beyond the critical frequency even of the F_2 layer, and will consequently penetrate the ionosphere and escape into space.

However, as might be expected considering the physical mechanisms by which it is produced, the ionosphere is a far from stable propagation medium, even at HF. There are many causes of variation, some regular and others irregular. As already noted, in addition to the diurnal variation, ionization in winter is less than in summer, and there is also a longer term variation due to cyclic change in the Sun's activity, usually called the **sunspot cycle** because it corresponds with the variation in the number of observable dark 'spots' on the Sun's disc. This cycle has an 11-year period.

There can be particular problems at high latitudes. In winter, during the polar night, electrons become so scarce that even HF quanta escape through the ionosphere, and the result is a **polar-cap blackout** of HF communication at high latitudes. Radio interruption also occurs when charged particles from the Sun, thrown out in solar flares (eruptions on the Sun which emit large amounts of radiation and ionized particles), are guided by the Earth's magnetic field into the polar ionosphere, creating F layer-like electron densities in the D layer. As a result HF can no longer pass through the D layer and polar-cap absorption of signals results. Such events are most likely near peaks of the sunspot cycle, and can occasionally be powerful enough to have much more extensive effects, greatly strengthening the absorbing D layer over a wide area of the Earth's surface and therefore raising the frequencies at which its absorbing effect is apparent far above the normal MF values. This constitutes a **sudden ionospheric disturbance** (**SID**), and may occur at any time, causing a complete disappearance of the sky wave over much, perhaps all, of the HF band. It may last only minutes but can extend to hours. **Ionospheric storms** typically occur some 30 h after a SID and are caused by the residual effects of sun flare radiation on the ionosphere, once the D layer has become transparent again. They have an effect all around the world and principally reduce critical frequencies. Lower operating frequencies are thus least affected.

As a result of experiments with very high-level nuclear explosions, the ionosphere is known to collapse completely in the location of high-yield nuclear bursts, and takes some hours to recover. Exact details are classified, but in a full-scale nuclear war the effects could be severe and widespread. Military planners are obliged to take this into account in their scenarios for maximizing population and battlefield resource survival in a major nuclear exchange and the 'broken-back' warfare which would follow.

13.6 Multipath effects in ionospheric propagation

Multipath propagation (Fig. 13.14) occurs when there exist two or more different viable propagation routes, for example surface wave and sky wave (particularly at MF) or single and multi-hop sky wave paths (at HF). As always it greatly modifies the characteristics of the received signal. We begin by considering MF with sky wave and surface wave interfering over an idealized flat Earth. The case is exactly as analysed above for terrestrial multipath (see Section 12.2, earlier), except that h is no longer small compared with d.

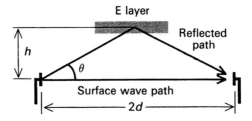

Fig. 13.14
Multipath transmission at MF by surface and reflected paths.

The sky wave path length minus surface wave path length difference is

$$\Delta = 2\left(\frac{h}{\sin\theta} - d\right)$$

so the phase difference is

$$\psi = \frac{4\pi}{\lambda}\left(\frac{h}{\sin\theta} - d\right) = \frac{4\pi f}{c}\left(\frac{h}{\sin\theta} - d\right) \quad (13.5)$$

The resultant of the two wave functions will pass through a zero when this phase difference is a multiple of π and maxima at multiples of 2π. Also, since h is not constant, approximately

$$\frac{d\psi}{dt} = \frac{4\pi f \cdot v}{c \sin\theta}$$

where

$$v = \frac{dh}{dt}$$

Thus the phase of the two waves varies continuously if the ionosphere moves up (or down) with a continuous velocity, such as in the evening when recombination is causing it to move higher.

The received signal passes through successive maxima and minima in a cyclic manner with a period T approximately given by

$$T = \frac{c}{2vf} \sin \theta \qquad (13.6)$$

Maxima occur when the phase difference is an even multiple of π and minima when an odd multiple; as the height of the ionospheric layer concerned changes the signal strength cycles through them, which is called **fading**. However, because frequency appears in eqn (13.6), at any moment of time, fading is different at different frequencies. This effect is known as **selective fading**. Note that when the path difference is short enough, selective fading will not be significant over the particular bandwidth of the transmissions in use. The term **flat fading** is then used to make this explicit.

Suppose that n is the number of wavelengths in the path difference at a frequency f_p where the received signal is at a maximum. The condition is

$$n = \frac{2\pi f_p}{c}\left(\frac{h}{\sin \theta} - d\right)$$

At minima

$$n + \frac{1}{2} = \frac{2\pi f_0}{c}\left(\frac{h}{\sin \theta} - d\right)$$

so

$$\Delta f = f_p - f_0 = \frac{f_p}{2n} = \frac{c}{2\Delta} \qquad (13.7)$$

Δf is, once again, the **coherence bandwidth**. This can be surprisingly small.

Selective fading spoils fringe reception of AM broadcasts in the MF band, particularly at night and in winter, due to sky wave interference with the normal surface wave service. As a result the AM carrier will go through zeros at times when there is sideband energy present and this will cause severe distortion, because the receiver demodulator normally depends on the presence of carrier. It also disables the receiver AGC (which is carrier related) so the distorted signal is also loud.

So far, the argument has been developed for the case of MF, where the multipath effect is between a surface and a sky wave propagation, but the analysis of selective fading at HF follows closely similar lines, except that in this case it is two (or more) sky waves which are interfering, whether they arrive by the alternative one-hop and two-hop paths or even propagate around the world in opposite directions (for example, east-about and west-about). Path lengths, and hence multipath differential delays, are an order of magnitude or two larger than in the MF case, and therefore n (the number of wavelengths in the path differential) is much larger. Despite the increased centre frequency, the result can be very low values for the coherence bandwidth. The effects can be very severe. As a result, HF transmissions are generally kept narrow band, to minimize the undesirable consequences of selective fading. SSB analogue speech transmission is in very widespread use, with an effective bandwidth of some 2.5 kHz. Data transmissions are generally restricted to low rate (formerly $\leq$ 2.4 kbits/s, although this is progressively improving), and the design of modems takes into account selective fading.

13.6 Ionospheric propagation: summing up

In the MF band, ionospheric propagation is simply a nuisance, leading to sky wave interference with the planned surface wave broadcast service provision, particularly at night and in winter, when the heavily absorbing D layer is not present. By contrast, it is sky wave propagation which plays the major role in the intercontinental communications capability of the HF bands, using the F layers. It is true, however, that HF channels have the reputation of requiring skilful management to give long-distance communication, which even then is of poor transmission quality, principally due to fading (Table 13.2). HF broadcasting, using AM, is particularly unsatisfactory over longer paths, because selective fading results in severe non-linear distortion in the receiver. The reasons for these problems are intrinsic to the mode of propagation of radio

Table 13.2 Establishing an HF connection

1. Determine the great circle distance between the stations
2. Depending on the distance and ionospheric layer to be used, determine the number of hops and reflection points
3. Determine mufs at reflection points from ionospheric data and set the transmission frequency near 85% of the lowest muf
4. Determine the elevation angle required at antennas
5. Compute the transmitter power required assuming an inverse square law, and adding a contingency margin of at least 10 dB
6. Try it, but be prepared to vary between multiple and single hops or even to send the signal the other way round the world if the mufs are more favourable

quanta via the ionosphere. As a result, in the 1960s and 1970s the use of HF declined somewhat, with interest turning to other intercontinental transmission technologies, particularly the use of satellites, despite the higher cost of these technologies.

However, it remains true that with careful selection of the HF transmission frequency, willingness to change frequency as required, and careful management of the antenna characteristics, it is virtually certain that satisfactory signal transfer on transcontinental contacts can occur, often without selective fading. In the past, doing this successfully depended on employing a highly skilled operator, who relied on regularly published ionospheric data along with much experience of past use of the band. Today, as a result of the rapidly falling cost of computing, 'intelligent' receiving equipment is available to do the task in return for a moderate investment. New self-optimizing HF systems either measure the characteristics of the ionosphere instantaneously by 'probe' transmissions (somewhat after the style of meteor scatter), or obtain similar information by measuring the characteristics of known regular broadcast transmissions from various distant sites. Improved signal processing and error correction (particularly using ARQ) has resulted in a better quality of demodulated signal, and new adaptive modems have raised available data rates, to the point where digital speech transmission is economic. Together these developments have resulted in a resurgence of interest in the HF bands at the present time, since it remains a very economic and flexible means of establishing intercontinental communications.

13.7 Satellite communications

The most important innovation in radio engineering of the second half of the twentieth century is communication with the aid of artificial satellites (Maral and Bousquet, 1998). The artificial satellite must be placed in orbit around the Earth. This is achieved when the object is given a velocity such that the gravitational attraction to Earth is balanced by forces generated by the curvature of its path combined with the velocity of its motion. The orbits are elliptical, although in many cases the ellipticity is so slight that they are very close to circular. As the altitude of the satellite increases the gravitational force lessens so its stable velocity decreases and the time it takes to circle the Earth (its **period**) increases.

None of the means of returning the radio energy directed into space described so far is without problems, in particular, both ionospheric propagation and meteor scatter have low data transmission rates compared with modern needs. In the late 1940s attempts were made to obtain signal return by reflection from the Moon, but it is simply too far away to give a satisfactory power budget in this service. Arthur C. Clarke first proposed the use of an artificial satellite carrying a radio transponder in the October 1945 issue of *Wireless World*. A low orbit satellite, *Sputnik 1*, was launched by the former Soviet Union on 4 October 1957, and created a worldwide sensation. The USA sent its own satellite into orbit some three months later. In 1963, *Syncom 2* was launched, the first synchronous satellite (its period matching the Earth's rotation). Since then, more than 3000 satellites have been successfully established in orbit. At first the overwhelming majority were constructed by the USA or former Soviet Union, but the European Space Agency (ESA) is now very active in space engineering. Canada, China, India and Japan have also launched satellites.

All artificial satellites tend to have certain features in common.

Solar cells generate the electrical power they use from the Sun, and storage batteries, recharged by the solar cells, provide back-up power when the solar light is obscured. In a very few cases satellites have used nuclear power sources. They also carry compressed gas and, indeed, exhaustion of stored compressed gas is a common life determinant for satellites. Controllable intermittent jets of gas are used to adjust the satellite's position periodically, offsetting orbital perturbations, and for attitude control equipment which keeps the satellite antennas pointed at the Earth target, using either the Sun, the edge of the Earth or a radio beacon on Earth as a reference point. Communications satellites carry all the necessary equipment to receive signals from the Earth stations, then re-transmit them with sufficient power to reach Earth again. Telemetry encoders measure voltages, currents, temperatures and other parameters monitoring the health of the satellite and send this information to Earth by radio. Finally, some kind of structure must house all this equipment. The weight to be contained is considerable (for example, an *Intelsat V* communications satellite weighed nearly 2 tonnes).

All such satellite systems must operate at radio frequencies high enough for the ionosphere to be penetrated (that is, at VHF or above). Even so, particularly at lower frequencies, there is both refraction and absorption of the radio energy. This is worst when the satellite is near the horizon and the wave propagation is at a very oblique angle to the ionospheric layers. Under these conditions, at VHF there may be total loss of signal. By contrast, above about 1 GHz all ionosphere effects vanish, so the VHF and lower UHF bands are now scarcely used for satellites.

To a first approximation at least, the mechanics of artificial satellites are simple. The gravitational force between the Earth (mass M) and a satellite of mass m distant r from the Earth's centre is

$$F = \mu \frac{m}{r^2} \text{ where } \mu = \frac{M}{\gamma}$$

where γ = the gravitational constant.

At the same time, if it can be assumed that the orbit is circular and that ω is the angular velocity of the satellite in radians/s, then the centrifugal force is F^*, where

$$F^* = mr\omega^2$$

For stability in orbit $F = F^*$, also the period $T = 2\pi/\omega$, so

$$T = \left(\frac{2\pi}{\mu^{1/2}}\right) r^{3/2} \tag{13.8}$$

But we can replace r by $(a + h)$, where a is the Earth's radius and h is the altitude of the satellite above the Earth's surface, so

$$T = \frac{2\pi a^{3/2}}{\mu^{1/2}} \left(1 + \frac{h}{a}\right)^{3/2}$$

The constant μ is equal to 3.986×10^{14} MKS units and a is approximately 6378 km, so

$$T = 5.071 \times \left(1 + \frac{h}{6.378 \times 10^6}\right)^{3/2} \times 10^3 \tag{13.9}$$

This relationship is plotted in Fig. 13.15. When the altitude is 35 786 km the period of the satellite is 24 h, which is the **synchronous** orbit. However, if the orbit is inclined to the equatorial plane there will still be some movement of the satellite as seen from Earth; a slight tilt, for example, will result in an apparent north–south oscillatory movement. By contrast, if the satellite is positioned in an orbit directly over the equator it will move exactly in step with the part of the Earth's surface beneath it, so that to an observer on the ground it will appear to hang stationary in the sky. This is a **geostationary** orbit.

13.8 Geostationary satellites

Because their capacity, in terms of the throughput of data, is determined solely by their on-board hardware, communications

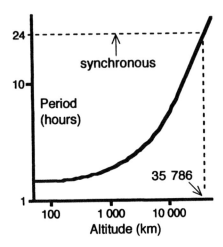

Fig. 13.15
Period versus altitude for an Earth satellite.

satellites have had a revolutionary impact on the practice of telecommunications. A measure of progress is the evolution from *Intelsat I* (1965) to *Intelsat VI* (1989). The earlier satellite weighed 68 kg and handled 480 telephone channels at a cost of $3.71 per channel hour, and had a working life of 1.5 years. By contrast, *Intelsat VI* weighed 3750 kg at launch and provides 80 000 channels, each at a cost of 4.4 cents per hour.

For use as a radio relay, a satellite in a geostationary orbit (**GEO**) is particularly useful because ground-based antennas with a very narrow main lobe, and thus high gain, can be pointed at the satellite without needing subsequent realignment. Similarly, antennas on the satellite can be permanently aligned on fixed targets on Earth. This ability to use high-gain antennas is essential because the principal disadvantage of geostationary satellites is their distance.

All GEOs receive the 'up' signal, increase its power, translate it to another frequency and re-radiate it back to Earth. Bearing in mind that the gain from the received signal to the transmitter output may be well over 100 dB, the frequency translation is essential because otherwise it would be impossible to prevent the powerful 'down'

Table 13.3 Principal frequency allotments for GEOs

VHF obsolete

UHF
Mobile service: 1.6 GHz up, 1.5 GHz down

SHF
Fixed stations:
6.725–7.025 GHz up, 4.5–4.8 GHz down
12.15–13.25 up, 10.7–10.95 and 11.2–11.45 GHz down

EHF/SHF
Around 30 GHz up, 20 GHz down

Also military allotments

---

signals from leaking into the 'up' receiving antennas, causing unwanted feedback and system malfunction. Frequencies are allotted (Table 13.3) by international agreement at the periodic World Administrative Radio Conferences (WARC) held under the auspices of the International Telecommunications Union, the world's oldest international body. So also are angular segments of the geostationary orbit committed to particular national administrations and suitably located for the territories concerned.

A serious problem for geostationary satellites arises from **orbital congestion**. All GEOs must be in the same equatorial orbit, which therefore becomes very congested over the more heavily populated longitudes. If the same up-frequency were used for all satellites, the only way of directing signals from Earth towards one rather than another would be by exploiting the directivity of the ground station antenna. This has severe limitations, however, because of the distance from Earth to GEO. This is so large that every 1° of main lobe width at the ground station corresponds to about a 600 km segment of orbit. It is hardly surprising that all the orbital locations serving Europe and the Americas were soon taken up. Since there are limits to the antenna directivity which can be engineered economically, the obvious solution to this problem is to operate different satellites at different frequencies, so that they do not interfere with each other.

The EHF band is particularly attractive for satellite use, since EHF antennas can be highly directive without being too large. For a paraboloid or an array, for example, the gain and hence the directivity is proportional to the square of a characteristic dimension expressed in wavelengths, so for a fixed lobe width the dimensions of the antenna vary directly with the wavelength. Thus going from 7 GHz (SHF, $\lambda = 4.2$ cm) to 30 GHz (EHF, $\lambda = 1$ cm) results in a reduction of size of the antenna by a factor of 4.2 (linear). Problems of atmospheric absorption are not too harmful at EHF because of the high elevation angle of the ground station antenna, which means that the radio quanta quickly pass out of the atmosphere. The only real disadvantages of EHF in this service have long been that a watt of EHF power was considerably more expensive than a watt of SHF and also EHF receivers were a few decibels less sensitive. Both of these problems are being progressively overcome. The use of EHF is therefore growing, and the problems of orbital congestion are being held off, for the time being.

There are two distinct types of GEO. A **transparent** satellite retains the received signal in radio frequency form, simply amplifying it and changing its frequency as may be required by mixing with a local oscillator. By contrast, a regenerative satellite demodulates the received signals, uses digital signal processing to re-shape them at base-band and then re-modulates them onto a new radio carrier. The latter are more complicated but able to return a better conditioned signal to Earth, eliminating most of the impairments of the 'up' transmission.

We have already seen that many of the problems of using geostationary satellites arise from the relatively great distance from the Earth of the synchronous orbit. So how far away from an Earth station is the satellite? From station at a latitude α to a GEO on its own longitude, as shown in Fig. 13.16, the distance is

$$z = a\left[1 + \left(1 + \frac{h}{a}\right)^2 - 2\left(1 + \frac{h}{a}\right)\cos\alpha\right]^{1/2}$$

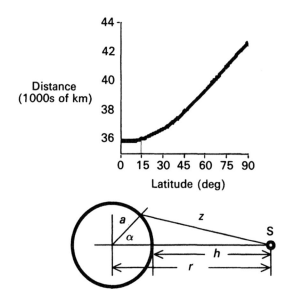

Fig. 13.16
Distance to the satellite as a function of latitude (not to scale).

For a GEO, we may take a as 6378 km and h as 35 786 km, so

$$z = 6378 \times (44.7 - 13.22 \cos \alpha)^{1/2}$$

With increasing latitude the distance of an Earth station to the satellite increases from the equatorial value of 35 786 km to a polar 42 669 km. Inverse square quanta spreading means that to a receiving antenna of 1 sq m aperture on the ground, at the equator the path loss from the satellite is 162 dB, and the 'up' path loss is the same. At the poles the corresponding figure is greater, but only by about 1.5 dB. However, at these latitudes other problems will be more severe, as the angle of elevation of the antenna gets less and less. The radio path becomes increasingly susceptible to near-surface losses and blocking by mountains or buildings, an effect known as **shadowing**. Although for fixed stations it is often possible to site the antenna to overcome the problem, it presents serious difficulties for land mobile installations (for example, on vehicles used in city environments). Finally, for some applications, like real-time speech, it is a further disadvantage that the round trip to the satellite and back takes 0.24 s at the equator, rising to 0.27 s at the pole.

Satellite transmit powers are limited by the available energy sources. Because the path loss is high, large EIRPs are required at the ground stations, but at the satellite the available transmitter power is limited to what solar panels can provide (70–100 W/sq m). This implies the need for high-gain, large aperture antennas on both the satellite and the ground station.

On the ground, the choice of antenna is between paraboloid antennas, usually with a waveguide horn feed, or active adaptive arrays. The former were universal at one time, but the latter look increasingly attractive with the fall in cost of semiconductors, since the ability to steer nulls at sources of interference is particularly valuable. So far as paraboloids are concerned, simply placing the primary feed at the focus of the reflector 'dish' has disadvantages (Chapter 9) and increasingly designers are using an offset feed or the Cassegrain configuration. A side lobe suppression collar around the feed, made of radio absorbent material and cooled for low noise, will maximize antenna performance. Because of the very weak received signal at ground level, side lobes are particularly undesirable in satellite ground stations.

Using high-power and high-gain antennas is costly and results in large Earth stations. Some military geostationary systems have achieved portability at the expense of severely restricting the system bandwidth (for example, to teleprinter signals only). The reduction of bandwidth results in a proportionate reduction of receiver equivalent input noise power, and hence makes a much smaller received signal power acceptable. This may meet a specific military need, but it loses one of the most important advantages of satellite circuits, which is that they can have very high data throughput rates, limited only by the radio bandwidth assigned to them and the capacity of the hardware.

Naturally, the antennas on the satellite are designed under much greater constraints than those of ground stations, since they must survive the launch and journey into orbit. Early satellites were not fully stabilized in spatial orientation, but were allowed to roll around one axis so that gyro forces would assist in holding their angle. This practice has been superseded by full three-axis stabilization, so that the vehicle may now be regarded as having a fixed orientation in space, which means that very narrow-beam antennas can be deployed. Paraboloid antennas are extensively used, unfolding after the satellite is established in orbit, as also are active arrays.

In Europe and many other areas of the world, high-power geostationary **direct broadcast satellites** (DBS) are commonplace, broadcasting to individual domestic TV installations. An early example was the Astra DBS satellite, which had 16 TV channels, with a power per channel of 45 W and a transmitting antenna gain of 35 dB. The EIRP in the central service area was +82 dBm (158.5 kW). Such a high EIRP made possible the use of small receiving antennas, typically 0.6 m in diameter in the principal service area and just a little larger in fringe areas. With the centre of its antenna 'footprint' located on the Franco-German border, the satellite gave satisfactory coverage over virtually the whole of the European Union, except for the extreme south of Italy. For services such as broadcasting, satellites are very economical of spectrum because they use only a single transmission frequency to cover a very wide area.

13.9 Low orbit satellites

Geostationary satellites do some tasks very well indeed but not others, whether by reason of their large path loss, poor Arctic and Antarctic coverage, or the long 'round trip' time delay. For this reason there is growing interest also in **low Earth orbit satellites** (LEOs). Often at only a few hundred kilometres altitude and frequently in polar orbits, they typically have periods of under 2 h (Fig. 13.15), and therefore any satellite will only be visible from a terrestrial site for a limited time, after which it will be obscured by the Earth until it 'rises' again on having completed an orbit. At any location, in short, they are only briefly (although frequently and predictably) receivable.

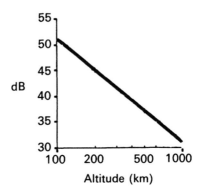

Fig. 13.17
Path loss advantage of an LEO over a GEO.

Low orbit satellites have a much more favourable power budget for radio links than GEOs, so terminal equipment could potentially be cheap (Fig. 13.17). They have two main disadvantages, however. The first is that the greater air resistance at low altitude gives rise to a shorter satellite life in orbit. Shorter satellite life can be countered by repeated launchings, however, and with the falling cost of rocket technology and the relatively unsophisticated LEO satellites mostly used, this additional space engineering cost is acceptable. The second problem, at the ground station, is that the movement of the satellite across the sky means that either the ground antenna must track it or have a wide enough lobe that the satellite is receivable for an acceptable period of time at each transit. This ground antenna problem presents some difficulties, because the cost of an automatic tracking antenna is generally prohibitive.

Early proposals concentrated on VHF satellites having large aperture ground-site antennas with a wide main lobe, typically dipoles or simple Yagi arrays, so that the need for the antenna to track the satellite is overcome. A simple store and forward capability in the satellite (in a low polar orbit) would give users the ability to communicate to any other site in the world, but only with short digital messages and with potential time delays up to an hour. Early systems like this aimed at very low cost; however, it is

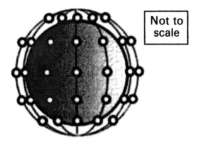

Fig. 13.18
The Iridium system of LEOs; 66 satellites in six polar orbits (not to scale).

arguable that an equally good service could be provided as cheaply by other means, such as HF radio, and few became operational.

Much more sophisticated low orbit systems are now being introduced, which show marked advantages against competing approaches. **Iridium**, launched commercially in 1998, is a system aimed initially at giving worldwide coverage to hand-held radiotelephones, but now offering a growing range of other global services (Fig. 13.18). It was designed by Motorola Inc. and is managed by an international consortium. As many as 72 satellites are each in polar orbit at 780 km in six orbit planes of 11 functioning satellites plus one spare, thus providing continuous line of sight coverage to any place on earth. The expected life in orbit of the satellites, each weighing 689 kg, is 5–8 years.

The satellites are equipped with several radio systems. Communication with the ground is at 1.616–1.6265 GHz for digital voice communication to hand-held terminals. The symbol rate transmitted is 2.4 kbaud and QPSK modulation is used. As well as digital speech, fax and data transmission are also supported. The main system ground stations communicate with the satellites at 29.1–29.3 GHz (up) and 19.4–19.6 GHz (down), whilst there are also 20 GHz links from each satellite to the one in front of it and the one behind in its particular orbit and to adjacent satellites in neighbouring orbits. Thus, each satellite is in communication with four others. As satellites pass over them, service users on the ground 'hand off' from one satellite to another in a way exactly analogous to the 'hand-off' by mobile users in an ordinary cellular scheme. In fact, the whole system has essentially the same mathematics as a cellular system, the only difference being scale and the fact that with Iridium it is the infrastructure which is moving.

Iridium is the first major and global commercial use of low orbit communications satellites. User terminals are already many and various, among them aircraft and ship installations, hand-held phones similar in size to cellular phones, solar-powered public call offices and a variety of data terminals. Between them they constitute a new worldwide and comprehensive telecommunications system, to complement and rival the established public network with which we are all familiar. Giving universal coverage without the need for a terrestrial infrastructure, the likely significance of this development is incalculable. See **http://www.iridium.com/** for further and up-to-date information.

13.10 Navigation satellites

Navigation satellites may be thought of as specialized low orbit satellites which provide the means to pinpoint any location on Earth with high accuracy by use of the Doppler effect. Once launched, the satellite's orbit is known precisely, which means that its velocity as well as its position is known at every instant of time. The satellite velocity can be accurately determined from an unknown position on the surface of the Earth by Doppler measurements made as it passes, as also can the time, from time signals that the satellite radiates.

The US **Transit** system has been in worldwide operation since 1964, used by more than a thousand stations, mostly merchant ships. Six satellites in polar orbit at 1100 km transmit at 150 and 400 MHz, broadcasting their position, with a time signal, at 2 min intervals. Users can determine their position 24 times each day to an accuracy of ± 175 m in latitude and longitude.

Navstar GPS (Global Positioning System) is much more advanced, consisting of 24 satellites positioned in six orbital planes of four each at 20 200 km altitude. The orbits are inclined at 55° to the equatorial plane and the satellites have a period of 12 h. They transmit on two frequencies: 1.57542 and 1.2276 GHz. The first Navstar GPS satellite was launched as early as 1978, and the full 24 production satellites were placed in orbit between 1989 and 1994. The service was formally inaugurated in December 1993, although it had been usable in part (with an earlier generation of satellites) for some years before that. Any point on Earth is always in view of at least five satellites and sometimes as many as eight.

The system is quite complex and will not be described in detail here. Full and up-to-date information can be obtained from **http://tycho.usno.navy.mil**. However, the upshot is that it provides users with their position in latitude and longitude accurate to 100 m, their altitude to 156 m and time to 340 ns at **standard accuracy**, all to 95% confidence. Standard accuracy Navstar receivers can be little larger or heavier than a cellular phone, and the most inexpensive cost about the same. In addition to standard there is also a **precise accuracy** service, available only to US government approved users, for which the corresponding figures are 22 m, 27.7 m and 200 ns. Finally, GPS can be used in a **differential** mode, to determine only the distance between two receivers. Because many of the sources of error are the same for both and therefore cancel, distances of up to 30 km can be determined with remarkable accuracy, to within millimetres with the best available equipment. This has revolutionized land surveying and geodesy.

There can be no doubt that the Navstar GPS system will transform many aspects of human life, in peace, and if need be, in war. Here too the implications will only appear with the passing decades.

13.11 The long haul – conclusions

For long-haul communications there are only four serious radio contenders: surface waves, meteor scatter, ionospheric propagation and satellites. The first was the pioneer long-distance radio technology, but its future is problematic. In the VLF band limiting rates of data transmission are very low, while the capital cost of installations is very high, due to the need for high-power transmitters and truly vast antenna systems. Because of the low data rate, VLF has not been seen as suited to speech transmission, but this may change due to the introduction of analysis–synthesis speech systems, in which computer-recognized words are signalled by short codes and synthesized into speech at the receiver. Present civil applications of VLF surface wave transmissions are principally for communicating time standards and in position-finding systems, both of which are now being taken over by satellite technology. There are niches, the most important being military, where the special characteristics of VLF surface wave transmission will enable it to survive, but there are not many. LF surface wave is a little better situated in respect of the ratio of capital cost to available data rates, but not much. Along with MF (itself not a long-haul band), the principal use of LF surface wave is for AM broadcasting and this will certainly continue for a long time to come, if only because of the enormous worldwide investment in receivers for this service.

Meteor scatter has interesting and unexpected characteristics. It gives only a very low transmission rate, some 200 bits/s averaged through the day, and it is strictly a point-to-point technology, useless for broadcasting. It cannot transmit real-time speech. However, it has very low capital cost and over distances up to about 2000 km it is beyond doubt the most reliable means of communication known. The equipment and antennas are compact and relatively invulnerable, while neither time of day, seasons, weather, sunspot cycles or even nuclear warfare would turn off this propagation mechanism. Indeed, only an unimaginable cataclysm of galactic proportions could do so, and humanity would not survive it. If you have only a little to say but must be sure that it will be heard, come what may, then meteor scatter at VHF could be your choice.

Ionospheric propagation, in the HF band, can transmit data at least two orders of magnitude faster than the average meteor scatter rate. It requires a modest capital investment, though a little higher than meteor scatter because of the cost of the larger antennas that are required for best performance. Analogue speech and even music transmission are within its capabilities, although the quality is often very poor, due to selective fading caused by multipath propagation, so its use for long-range broadcasting leaves much to be desired. In point-to-point applications, per channel data rates are just sufficient to sustain digital speech, using sophisticated codecs. At one time it was argued against this technology that it demanded skilled operators for effective use, but this is no longer the case as a result of computer control. However, long-distance HF circuits are subject to unpredictable interruptions arising from the character of the ionosphere itself, and the interruption or degradation of service may last from minutes to, in very rare cases, a day. From the military standpoint, in a major nuclear exchange the ionosphere would probably be unusable, taking many hours to recover. Even so, it would be restored long before the satellites. Perhaps the future of ionospheric propagation as a long-haul point–point carrier is for users who are mobile or remote from the telecommunications network, and as a last-ditch back-up to satellite communications. There can be no doubt that a major objective in any future war must be the destruction of the enemy's satellites.

War circumstances apart, unquestionably far and away the most important long-haul communications technology will exploit the characteristics of artificial satellites, for which the only real disadvantage is potential vulnerability. In peace-time, GEOs facilitate very high-capacity intercontinental communication, while also having a most important role in broadcasting. Although the capital investment required to set up a satellite system is very large, the capacity provided by even a single GEO is so great that the cost per bit of information transferred is very low indeed, provided the circuits can be kept fully loaded. Thus, if the traffic volume is sufficient, a GEO is far the most economic long-haul radio system, despite its high initial cost. Indeed the major economic competitor to the GEO is not a radio system at all but fibre optic cable, although this can only be used between fixed locations.

As for broadcasting, three technologies are important, and we may be sure that terrestrial transmissions, cable systems and direct broadcasting satellites will continue to be in contention for the foreseeable future. However, terrestrial broadcasting is very inefficient in spectrum use and, due to the high installed cost per kilometre of cable, cable systems are only viable in areas of substantial population density, say 100 persons per square kilometre or more. In many other parts of the world, where people are thin on the ground, only DBS could provide developed television services at reasonable cost.

No less exciting is the potential for LEOs, which again demand a high initial investment, but have the potential of working with very simple, cheap and portable ground terminals, because of the much lower path loss. In both communication and positioning systems their impact is already clear and dramatic. Many think that they spell the end of both VLF radio navigation systems and HF ionospheric systems for commercial point-to-point communication, leaving HF as an amateur, military and broadcasting band only. That view is perhaps still controversial but it may prove right.

To summarize, the present and future of long-haul radio communications must be predominantly with the satellites, whether GEOs or LEOs, and other technologies will survive essentially in specialized niches where their particular characteristics confer an advantage.

Problems

1. An AM broadcast transmission at 1 MHz reaches a receiver by both sky and surface wave at equal strength but with 50 km path length difference. What is the coherence bandwidth and what significance would you attach to this figure? [3.0 kHz]

2. A meteor scatter site is to be established solely to communicate with another at a distance of 1000 km. Meteor trails occur between about 20 and 50 km altitude. What antenna elevation would you use? [4°] What is the minimum width of the main lobe? [3.6°] What maximum antenna gain would you estimate? [33 dB]

3. An HF radio link is to be established between Dublin (Ireland) and Madras (India). Enquiries to operators in Baghdad (Iraq) confirm that satisfactory communication is being obtained with

both these cities at frequencies of up to 11 MHz using optimal antenna elevations of 10.4°. What is the muf for a single-hop Dublin–Madras circuit and what will be the optimal elevation angle? [21.7 MHz, 5.25°] (Assume that Baghdad is equidistant from Dublin and Madras on a line between them, and that Madras is 8710 km from Dublin.)

4. If an HF transmission at 3 MHz has 500 km path difference between one- and two-hop propagation modes, what is the coherence bandwidth? [300 Hz] What will be the effect on analogue speech transmission? And on data?

Appendix

Feeders

Feeders used with radio antennas are a particular type of **transmission line**. In the past, books about antennas often spent a large proportion of their pages on their theory and properties. However, in the last quarter of the twentieth century radio technology changed in ways that make this no longer a reasonable thing to do. In times gone by, antennas were often remote from the receiving or transmitting equipment, with the result that the feeder run was many wavelengths long. Modern electronics is more compact and consumes less power, so it can be packaged in or near the antenna, with the result that the high frequency feeder is either short or virtually non-existent. Even where this is not so and, for example, high-power MF transmitters are still remote from their antennas, it is now common practice to integrate a digitally controlled ATU with the antenna, so that the feeder can be operated in a matched mode (of which more later). Contemporary practice is overwhelmingly to design for feeders operating matched and as a result largely without standing waves. This, together with the shorter length, results in very little of the resonance phenomena with which an earlier generation of engineers battled.

With the arrival of radio frequency microelectronics there has been a general abandonment of distributed-constant circuits, which use lengths of transmission line as components, in favour of lumped-constant circuits. Since the shortest wavelengths of much interest to radio engineers exceed 1 mm, yet this is more than a thousand times the feature size on a microelectronic circuit, the trend to lumped circuits must intensify rather than reverse. Thus transmission lines operating under odd and mismatched conditions are no longer as interesting to radio engineers as they once were. Finally, few engineers now design their own transmission lines. These are either regarded as bought-in components, if discrete components, or they are laid down on microelectronic circuits in accordance with tight design protocols established by the semiconductor processor.

This Appendix will therefore summarize the properties of transmission lines operating in a matched mode, doing so only to the extent that is required for understanding earlier parts of the book. The simplest transmission line (Fig. A1) consists of two straight parallel wires. These are shown as two cylindrical conductors, rods or wires perhaps, of radius a separated by a distance between their centres D, but the theory is little changed if the conductors have any other shape in cross-section. We begin by imagining that they extend to infinity. If we connect a radio frequency generator between the ends of the line, radio energy will begin to propagate along it, dissipating by various loss mechanisms until far enough from the generator it falls below the noise level. No energy flows back and there are no standing waves. At the input terminals of the line there will be a certain current for any applied RF voltage, depending only on a, D and the properties of any insulating material around the conductors. The ratio of voltage to current at the input terminals of an infinite transmission line is called the **characteristic impedance** of the line and designated as Z_0.

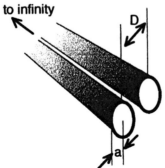

Fig. A1
The parallel line feeder.

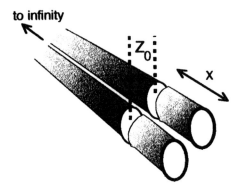

Fig. A2
Terminating the line in its characteristic impedance is equivalent to extending it infinitely.

If the line is now cut at some finite distance x from the input port (Fig. A2), the part of the line beyond the cut is still infinite and therefore still looks like Z_0 at its input terminals. It must follow that if we now terminate the length x of the line before the cut in an impedance Z_0 the conditions at the input port of this length will be exactly as they were before the cut: there will be no reflection, no standing waves and the impedance at the input port will still be Z_0. Under these conditions the finite section of line is said to be **matched**, and as near as possible this will be how it will be operated. Because terminating a transmission line in its characteristic impedance results in matched operation it is sometimes called the **matching impedance**. By avoiding standing waves we avoid resonance effects and hence undesirable frequency dependence of the line characteristics and we also reduce the maximum voltage on the line, which improves its power-handling performance.

For a parallel-wire transmission line of the type described at radio frequencies the characteristic impedance is given by

$$Z_0 = \sqrt{\frac{L}{C}} \qquad (A1)$$

where L is inductance and C capacitance per unit length.

From this general expression a value of Z_0 can be calculated, knowing the shape and dimensions of the conductors in a particular case. For a parallel-wire line with round conductors (Fig. A1)

$$Z_0 = \frac{120}{\sqrt{\varepsilon}} \log_e\left(\frac{D}{a}\right) \quad \text{(ohms, if } D \gg a) \tag{A2}$$

Here ε is the relative permittivity (dielectric constant) of any insulating material there may be between and around the conductors, while it is assumed that the surrounding medium has no special magnetic properties. If a dielectric is present the speed of the photons adjacent to the conductors is less than c by a factor dependent on the permittivity, thus the velocity in the transmission line (assumed totally immersed in the dielectric) is

$$v = \frac{c}{\sqrt{\varepsilon}} \tag{A3}$$

This figure can be important where lengths of line are used to create time delays, as in phase shifters for array antennas. For polythene, the most widely used RF insulating material, the relative permittivity is 2.2 so the velocity in a line insulated by this material is 67% of the free-space velocity of light, while for silicon dioxide (used in microelectronic circuits) the figure is 47%.

It has already been mentioned that transmission lines have losses, progressively dissipating the energy flowing along them. How does this come about? A crowd of radio quanta travels between and near the conductors with the speed of light. What is happening is that near-field quanta are emitted in large numbers and soon afterwards mostly recaptured, but a few escape altogether. Since those escaping represent a loss of energy, a good transmission line minimizes this. How is it done?

If the conductors are close the field between them is large, so the wave magnitude is also large, and that results in a high probability that the quanta will all be crowded between the conductors, where they have little chance of escape. Thus, if each conductor is well within the near–far transition radius of the other the chance of quanta escaping is minimized, whereas if they are far apart the

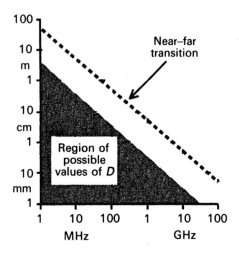

Fig. A3
Parallel line spacing falls with increasing frequency.

quanta increasingly move out from between the conductors and much energy will be lost. So D, the spacing between the conductors, must be small compared with the near–far transition distance, $\lambda/2\pi$, and is typically around one-hundredth of a wavelength (Fig. A3). At high frequencies D gets very small, and therefore a is much smaller still, making the conductors fragile and tricky to handle. This has two negative consequences: it may be difficult to fabricate the line and its power-handling capacity will be reduced due to the risk of insulation breakdown between the conductors.

Radiated quanta are not the only way that a transmission line can lose energy. There will also be the usual ohmic losses in the conductors, growing larger as the conductors get smaller, and the dielectric losses in the insulating material between the lines can be increasingly severe as the operating frequency rises. To summarize, all transmission lines have losses and these mostly increase with frequency. For example, at 50 MHz standard commercially available 50 Ω feeder cables have losses ranging from 0.015 dB/m to 0.1 dB/m, while at 450 MHz the same cables have losses of 0.07 dB/m and 0.5 dB/m, respectively.

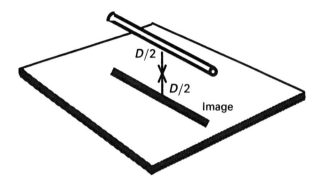

Fig. A4
An image in a conducting plane may form the second conductor.

An alternative transmission line has one wire over an infinite (in practice large) conducting plane (Fig. A4). This is not fundamentally different from the two-wire version, but as usual relies on image formation, and thus the distance between wire and plane is half that for a comparable two-wire line. Whereas the two-wire line is a **balanced** line, with the alternating voltage applied between the two terminals and neither conductor an equipotential, the single-wire line is **unbalanced**, in that the function of the image is to keep the potential on the conducting plane everywhere the same (see Section 6), so that it is an equipotential surface. Practically speaking this is an enormous advantage since it can be earthed (grounded), which means that it can be common with other components and can be attached to or even form part of the equipment structure if desired. Wherever possible unbalanced feeders are always preferred to balanced ones for these very good practical reasons.

With a monopole (Marconi) antenna, which is naturally unbalanced, the use of an unbalanced feeder presents no problem, but a dipole is balanced, so special arrangements have to be made. A **balun**, a balanced-to-unbalanced converter, is interposed between feeder and antenna in this case (Fig. A5). There are many types of balun, of which conceptually the simplest is just a transformer, although they can also be constructed from suitable lengths of unmatched resonant transmission line.

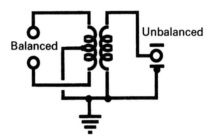

Fig. A5
A balun based on a transformer may also provide impedance transformation.

By far the most important type of transmission line is a variant of Fig. A4 in which the conducting plane is wrapped around the wire in the form of a tube. This is the very familiar coaxial transmission line, certainly the most widely used of all (Fig. A6). Because the conducting plane now entirely surrounds the inner conductor the photons' of radio energy are entirely trapped within the space between the inner and the coaxial sheath, so there is virtually no energy loss by radiation arising from their escape. Like the flat-plane variant from which it derives, it is an unbalanced transmission line, and the outer surface of the sheath can be earthed at any point. Normally made flexible with a braided outer conductor,

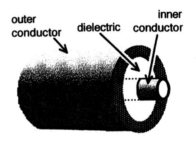

Fig. A6
If the plane in Fig. A4 is wrapped around the conductor a coaxial transmission line results.

polythene dielectric and a stranded inner core, it is available in many sizes and different characteristic impedance, although 50 Ω is by far the commonest value. At the other extreme, coaxial lines for the highest powers may be made with rigid copper tubes as inner and outer conductors, and the space between them filled with pressurized inert gas.

The expression for the characteristic impedance of a coaxial line is

$$Z_0 = \frac{60}{\sqrt{\varepsilon}} \log_e \left(\frac{R}{a}\right) \quad \text{(ohms)} \tag{A3}$$

where R is the outer conductor radius and a the inner.

Another very important type of transmission line has rectangular 'ribbon' type conductors in place of wires or rods. All such lines are called **striplines**, and there are very many different versions (Fig. A7). Perhaps the simplest is the ribbon cable, which has two strip conductors separated by a dielectric, usually polythene (Fig. 8.7(a)). Since the ratio of capacitance to inductance per unit length is large, it is useful when a cable of very low characteristic impedance is required (see eqn (A1)).

More significant is the version using a ribbon conductor over a plane. This is very widely used in MICs and MMICs where transmission lines are required, as well as on printed circuit boards. The ground plane may be a copper foil in the latter case, with a

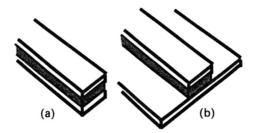

Fig. A7
Rectangular conductors replace wires. There are many variants of this stripline, from (a) ribbon cables to (b) microstrip.

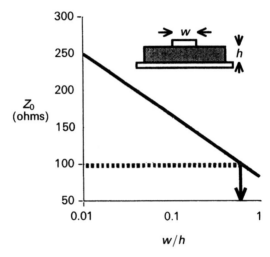

Fig. A8
Characteristic impedance of a microstrip transmission line.

plastic dielectric, whilst on integrated circuits it is either an evaporated metal film or a heavily doped semiconductor substrate, and the insulating layer is usually silicon oxide or a glass.

For **microstrip** lines of this kind the characteristic impedance is as shown in Fig. A8, which assumes that the dielectric is silicon dioxide with a relative permittivity just over 4. If some other dielectric were used the characteristic impedance would vary inversely as the square root of the permittivity (eqn (A1)). Over recent years, 100 Ω has become a *de facto* standard for transmission lines on and between MMICs. For microstrip lines insulated with silicon dioxide this happens when the width of the conducting track is 0.6 of the thickness of the insulating layer.

As already noted, the losses of all the types of feeders described so far increase rapidly with frequency. Although cables specially designed for higher frequencies can improve matters somewhat, the problem becomes increasingly difficult into the microwave region. During World War II (1939–45) in the earliest stage of the use of microwave radio technology for radar, when equipment was much bulkier and consumed far more power than today, it was often necessary to site transmitters and receivers far from their antennas, and the need for some lower loss form of feeder was consequently pressing. The first attempted solution was to use coaxial feeders insulated almost entirely by air, which virtually eliminated dielectric losses, but as frequencies were still further increased this proved inadequate.

Since microwaves penetrate only a very short distance into conductors (**skin effect**) it was obvious that the residual losses could be minimized by reducing the internal surface area of the coaxial feeder, and the most radical way to do this was to remove the central conductor altogether, so that the feeder became no more than a conducting tube. Thus was born the concept of a **waveguide**, which has far lower high-frequency losses than the conventional coaxial cable. Waveguides are metal tubes, usually rectangular in section, in which photons are launched at one end by means of a short monopole or a magnetic doublet. Photons then make successive reflections from the walls as they progress along it. There are a number of patterns in which this can happen, called the **modes** of the guide.

In recent years the use of waveguides has declined, partly because of the high cost of waveguide components fabricated from metal to very close tolerances, but also because changes in technology have made waveguides less necessary. Modern receivers often use a down-converter sited very close to the antenna, so that the subsequent cable run is at relatively low frequency. As for transmitters, they are increasingly being integrated with antennas, as active arrays, or else are sited at mast-top, close to the antenna. Today, only the highest power transmitters are now situated far from their antennas, and therefore need long feeder runs. A detailed description of waveguide technology would be out of place here; there are already many excellent texts on the subject (Baden Fuller, 1990).

Further reading

Allsebrook, K. and Parsons, D. (1977) Mobile radio propagation in British cities at frequencies in the VHF and UHF bands, *IEEE Trans. Veh. Tech.*, **VT26**.
Baden Fuller, A.J. (1990) *Microwaves*, 3rd edn. Pergamon, Oxford.
Connor, F.R. (1989) *Antennas*, 2nd edn. Arnold, London.
Delogne, P. (1982) *Leaky Feeder and Subsurface Radio Communication*. Peregrinus, Stevenage.
Egli, J.J. (1945) Radio propagation above 40 Mc over irregular terrain, *Proc. IRE*, **45**.
Gosling, W. (1978) A simple mathematical model of co-channel and adjacent channel interference in land mobile radio, *Radio Electronic Eng.* **48**.
Isbell, D.E. (1960) Log-periodic dipole arrays, *IRE Trans. Antennas Propagation*, **8**.
Jakes, W.C. (1974) *Microwave Mobile Communications*. Wiley, Chichester.
Kraus, J.D. (1989) *Antennas*. McGraw-Hill, New York.
Kraus, J.D. (1992) *Electromagnetics*. McGraw-Hill, New York.
Lee, W.C.Y. (1989) *Mobile Cellular Telecommunications Systems*. McGraw Hill.
Maral, G. and Bousquet, M. (1997) *Satellite Communications Systems*, 3rd edn. Wiley, Chichester.
Orr, W.I. (1987) *Radio Handbook*, 23rd edn. Sams, Indianapolis (reprinted 1995).
Rumsey, V.H. (1957) Frequency independent antennas, *IRE Nat. Conv. Record* Part I.
Torrance, T.F. (Ed.) (1982) *James Clerk Maxwell: A Dynamical Theory of the Electromagnetic Field*. Scottish Academic Press, Edinburgh.
Turkmani, A.M.D., Parsons, J.D. and Lewis, D.G. (1987) Radio propagation into buildings at 441, 900 and 1400 MHz, *Proc. IERE Fourth Int. Conf. on Land Mobile Radio*, **78**.

Index

Absorption 128, 130, 151, 166, 172–4, 198, 206, 213, 218, 220, 228, 232
Aether 6, 10
Air traffic control 1, 65, 159
Ambulance 1
Ampère, André Marie 5
Antenna:
  adaptive/active array 61–70, 97, 105, 109, 114, 121, 141–3, 146, 158, 234, 253
  aperture 24, 35, 40, 47–9, 55, 56, 60, 76, 77, 79, 85, 95, 114, 126, 130, 137–48, 156–7, 208, 233, 236
  array 52–84, 89, 92, 101, 106–17, 121, 125, 131, 140–4, 146, 149, 158, 179, 208, 230, 235, 247
  efficiency 19, 45–6, 49, 75, 101, 104, 110, 203
  equivalent circuit 41–50, 72, 74, 75, 117, 118
  gain 33–5, 66, 108, 112, 126, 141–2, 157, 234
  horn 144–5
  log periodic 106–9
  long-wire 110–14
  loop antenna 116–20, 128
  magnetic 116–17
  Marconi 98, 203, 249
  microwave 133–49
  omnidirectional 58, 98, 102, 109, 139, 142
  parabolic trough 90–93
  paraboloid 94–7, 143–8, 153, 158, 232, 234
  'smart' 62
  tuning unit (ATU) 44–5, 47, 100, 103, 105–6, 117–19, 127, 203, 244
Arago, Jean 4
Armstrong, Edwin 205

Broadcasting 1, 59, 79, 89, 96, 100, 104, 121, 128, 135, 139, 163, 170, 202, 205, 221, 225, 235, 240–2
Broadside 32, 47, 54, 59, 126
Buildings 129, 151, 176–8, 187, 190, 196, 198–200, 233
  techniques 199
  penetration 198–9

Capital investment 240–1
Cardioid 55, 61
Cellular radio 2, 196, 197, 198, 202, 236, 238
Classical mechanics 2, 19
Coaxial 128, 130, 134, 250, 252

Collision 172, 173, 206, 213, 214, 215, 218
  cross section 92, 94, 161, 164, 213
Computer 1, 102, 152, 211, 240, 241
Conducting surface 81–102, 128, 203
Contour map 168
Corner reflector 86–7
Correspondence principle 11
Critical frequency 215–22

Data rate 204, 205, 226, 239–41
Decibel notation 28–30
Diffraction 4, 10, 86, 151, 169, 177, 182–9, 198, 200, 204
  Fraunhofer 183
  Fresnel 183
Digital speech 226, 237, 240, 241
Dipole 26–50, 60, 64, 72, 76, 77, 81, 83–7, 89, 91, 94, 104–8, 116–17, 120, 137, 139–42, 145, 153, 236, 249
Director 75–80, 84, 107
Doublet 26, 31, 116, 127, 128, 253
Doughnut 31, 47, 52, 56, 61, 78, 86, 99, 117, 153
Duals 116
Ducting 170–2, 179

Earth 96, 151, 152, 169, 177, 207, 208, 215, 218, 228
  atmosphere 152, 165–76
  stations 228
  surface 139, 158, 166, 177, 179, 203–5, 206, 207, 212, 221
EHF 15, 120, 133, 134, 136, 148, 172, 174, 175, 179, 188, 199, 231, 232
EIRP 154–6, 160, 174, 232, 234
Electromagnetic spectrum 13, 14, 127
  waves 9, 10, 12, 14
Electronic warfare 174

Elint 174
ELF 14, 46, 179, 187
End-fire 55, 58, 60, 74, 76, 106, 111, 121, 126
ESM 164
Exponential distribution 209

Fading 182, 190, 194, 224
  flat 224
  Rayleigh 190–4
  selective 224, 225, 241
Far field 23, 27, 30, 52, 127, 129, 151, 153
Faraday, Michael 4–6
Ferrite 118–20, 203
Fessenden, Reginald 205
Focus 92, 146, 147, 234
Footprint 139, 208, 235
Fourier spectrum 159
Frequency assignment 200
Fresnel, Augustin 4, 183

GEOs (geostationary sat's) 229–35, 241, 242
GPS (global positioning system) 238–9
Gravitation 2, 153, 225, 227, 228
Gravitons 2

Heaviside, Oliver 166, 212
Helical antenna 121–7
Henry, Joseph 4
Hertz, Heinrich 10–13, 167
HF 14, 40, 56, 64, 79, 99, 102, 104, 105, 108, 109, 114, 119, 131, 152, 171, 179, 187, 189, 204, 205, 212–27, 237, 240–3
High power 20, 100, 134, 137, 144, 204, 206, 234, 235, 240, 244
Hop, multiple 218, 221
Huygens, Christiaan 3

Index

Illuminator 162–3
Image 81–5, 87, 93, 99, 110, 128, 183, 249
Imaging 1, 81, 175
Induction 24, 28, 73, 76, 119
Infra-red 13, 172, 200
Interference 10, 67–70, 79, 97, 102, 109, 119, 129, 142, 147, 151, 171, 172, 174, 181, 196, 208, 218, 224–5, 234
Ionized layer 214, 216
Ionosphere 166, 179, 204, 212–27, 240–2
Ionospheric propagation 104, 152, 206, 212–27, 240–1
Ions 166, 207, 213
Isotrope 20–5, 28, 33, 34, 61, 95, 153–4
Isotropic 20–8, 30–4, 48, 55, 61, 85, 95, 108, 126, 137, 141, 142, 144, 151, 152–4

Jamming 67, 70, 79, 97, 103, 147, 159

Kennelly, Arthur 166

Lambert's cosine law 178
Latitude 96, 139, 171, 209, 221, 232, 233, 238–9
LEOs 152, 235–8, 242
LF 15, 100, 118–19, 128, 151–2, 179, 187, 204–5, 221, 240
Light 2–4, 9, 13, 182–3
Lobe steering 63–6, 70, 79, 106, 142
Long-haul 202–42
Longitude 230, 232, 238–9

Magnetic poles 6
Main lobe 56–8, 59, 60–9, 75–8, 85–90, 93–6, 102, 106, 111, 113, 114, 122, 139–41, 142, 145, 154–5, 158, 206, 208, 230, 236
Marconi, Guglielmo 98, 100, 166, 204
Mast 50, 79, 99, 100, 101, 108, 110, 113, 114, 115, 121, 168, 169, 170, 171
Maxwell, James Clark 4, 5, 6, 7, 8, 9, 10, 11, 12, 13, 19, 23, 26, 27, 34, 38, 81, 82, 149, 189
Mesosphere 166
Meteor 207
  scatter 207, 209, 210, 211, 225, 226, 239, 240, 242
  systems 208
  trails 179, 210
MF 14, 79, 97, 99–102, 109, 118–19, 128, 151, 179, 187, 203, 205, 220–1, 222–5, 240, 244
Microwave 133–49, 167, 177, 252–3
  technology 133–7
Military 1, 67, 70, 79, 103, 105, 121, 147, 152, 161–3, 174–5, 198, 199, 202, 204–5, 209, 221, 234, 240–2
Millimetre waves 133, 188, 199, 200
Mobile phone 89, 193
Modes 151, 199, 253
Momentum 11
Mountain 177, 179, 181, 187, 233

Navigation 1, 204, 238–9, 242
Navstar GPS 238–9
Near field 23–4, 26, 28, 72, 119, 127–31
Near–far transition 23, 27, 110, 119, 248
Newton, Isaac 3, 183
  laws 2
Nuclear exchange 222, 241
Null steering 68–70, 79

Oërsted, Hans Christian 4

258 Index

Operating frequency 44, 51, 74, 105, 113, 117, 249
Orbit 139, 152, 207, 227–39
Orbital congestion 231–2
Ozone layer 165–6

Parasitic elements 72–6, 81, 102
Pencil beam 64, 94, 139, 153, 158
Period 8, 41, 160, 210, 221, 224, 226, 227–39
Phase shifter 63–4, 247
Photon 2–4 and many subsequently
Poisson, Simeon 4
Polar diagram 28, 30–5, 48, 50, 54–61, 67–8, 75–9, 81, 84, 86, 93, 96, 99, 102, 106, 108, 110, 112, 117, 119, 141, 142, 153, 154
Polarization 34–8, 93, 113, 117, 121, 127
Pompadour, Madame de 4
Power:
  amplifier 64–5
  devices 159
  gain 33–5, 47–8, 52–5, 56–8, 59, 61, 95–6, 108, 112, 126, 141, 154–5, 157
Pulse modification 159

Quantum 2–4 and many subsequently
Quantum mechanics 2, 10–13, 183

Radar 1, 56, 65, 66, 70, 79, 93, 96–8, 103, 134, 135, 139, 148, 151, 158, 158–63, 170, 175, 252
  absorbent material 162
  bistatic 162
  continuous wave (CW) 160
  cross section 160–2
  monostatic 162
  multistatic 162–3

Radio:
  coverage 97
  receivers, reception 1, 18 and many subsequently
  regulatory authorities 200
  systems 1, 2, 12, 179, 182, 196, 208, 237
  transmitters, transmission 1, 18 and many subsequently
Rayleigh, Lord (J. W. Strutt) 190
Rayleigh distribution 130, 190–9
Recombination 207, 210, 212–16, 220, 223
Reflected 40, 41, 85–6, 93, 97, 99, 130, 138, 177, 179–80, 190, 213, 217, 218, 221
Reflection 10, 84, 87, 90, 109, 110, 111, 114, 129, 135, 145, 149, 151, 152, 160, 162, 164, 177, 178, 179, 180, 181, 189, 193, 196, 199, 216, 217, 218, 226, 246, 253
  coefficient 178
  diffuse 178
  specular 178, 218
Reflector 64, 74–6, 77, 87–96, 107, 138, 143–8, 149, 153, 157, 177–9, 196, 208, 218, 234
Refraction 10, 149, 166–72, 189, 213, 217, 228
Refractive index 141, 148, 167, 170
Resistance:
  loss 42–6, 49
  matching 45, 50
  radiation 42–6, 49–50, 74, 105
Resonance, resonant 41, 42–5, 49, 73–5, 100, 104–5, 107, 109, 113, 114, 119, 144, 198, 203, 244, 246
  frequency 41, 42, 44, 103, 107
  length 44, 74, 76, 100, 105
Room to room propagation 200

Index **259**

Satellite 96, 139, 152, 175, 206, 211, 213, 224, 226, 227–39, 241–2
Scattering 146, 189–98, 200, 206
Shadow 152, 177, 182–9, 200, 233
SHF 14, 79, 113, 121, 126, 133, 134, 148, 179, 199, 231
Sky wave 108, 171, 205–6, 207, 220–6
Sigint 174
Silicon 136, 142, 247, 252
Slot antenna 120–1
Spectrum conservation 200
Standing wave 41, 50, 109–10, 134, 144, 244–6
Stealth 160–1
Stratopause 166
Stratosphere 164–5
Surface wave 151–2, 203–5, 217, 220–1, 225, 239

Tanks 1, 161, 175
Telemetry 228
Television 1, 77, 85, 104, 108, 121, 135, 137, 145, 163, 172, 242
Tesla, Nicola 212
Thermosphere 166
Transit 238
Transmission line 18, 63, 134–6, 145, 244–53

Trees 179, 196–7
Tropopause 165, 166, 208
Troposphere 165, 166, 170, 206

UHF 14, 40, 46, 75, 77, 79, 99, 108, 114, 126, 129, 133, 171, 172, 179, 187, 190, 195, 196–8, 199, 206, 221, 228
Ultraviolet 13, 166

Velocity of light, c 3, 8–9, 21, 49, 166–7, 247
VHF 14, 46, 59, 75, 79, 104, 109, 119, 126, 129, 152, 168, 171–2, 177, 179, 187, 189, 190, 196, 197, 199, 205, 206, 207, 208, 219, 228, 236, 240
VLF 14, 101, 152, 179, 187, 203, 204, 205, 239, 240, 242

Waveguide 134–5, 144, 147–8, 149, 234, 253
Whip 99, 102, 105, 127, 205

X-ray 13, 212

CPSIA information can be obtained at www.ICGtesting.com